INTERNATIONAL CENTRE FOR MECHANICAL SCIENCES

COURSES AND LECTURES - No. 26

MAREK SOKOLOWSKI
POLISH ACADEMY OF SCIENCES

THEORY OF COUPLE-STRESSES IN BODIES WITH CONSTRAINED ROTATIONS

COURSE HELD AT THE DEPARTMENT
FOR MECHANICS OF DEFORMABLE BODIES
JULY 1970

UDINE 1970

SPRINGER-VERLAG WIEN GMBH

Originally published by Springer-Verlag Wien New York in 1972

ISBN 978-3-211-81143-6 ISBN 978-3-7091-2943-2 (eBook)

DOI 10.1007/978-3-7091-2943-2

P R E F A C E

These notes represent the material of the author's lectures given at the CISM summer courses in Udine 1970.

The author is indebted to Prof. Luigi Sobrero, Secretary General of CISM for his kind invitation to deliver these lectures. As always, the staff of CISM should be thanked for the trouble they take in printing, proofreading and publishing these lectures.

Udine, July 1970

M. Sokolowsky.

1. Introduction

In the last ten, twelve years a rapid development of the theory of Cosserat-type media, bodies with additional internal degrees of freedom, the couple-stress theory and the theory of bodies with internal microstructure can be observed. It is not the aim of this series of lectures to characterize generally the above mentioned theories, to outline their fundamental physical and mathematical assumptions and to discuss the possibilities of their practical applications in numerous fields of technology. Some of these problems are subject to detailed discussion in the lecture to be held in the same period of time at Udine by Professor R. Stojanovic and Professor W. Nowacki; some of these problems are not completely cleared in the world literature so far- like, for instance, the extent of possible technological applications of the couple-stress theory to real polycrystalline bodies.

Generally speaking, two trends can be observed in the current development of the theories mentioned above; one of them consists in a constant broadening of their mathematical and physical basis; increasing the number of internal degrees of freedom, introducing a number of new elastic and material constants; trying to embrace the largest possible number of physical phenomena and to generalize most of the results known from the classical theories of deformable bodies.

The other tendency can be characterized by the efforts to obtain certain new solutions in this domain based, however, on possibly simple models of media, and on the lowest possible number of additional assumptions, material constants, etc., in order to obtain solutions which could successfully be compared with certain experimental results which cannot be explained on the basis of classical theories of elastic media. The so-called theory of bodies with constrained rotations offers a good opportunity for this type of investigations. Under the assumption of isotropy, centrosymmetry and homogeneity, the number of additional material constants is reduced to two (or even one, depending on certain additional assumptions); a series of problems of well-known classical problems of elasticity can be solved, within the framework of this generalized concept of elastic medium, in a more or less simple manner; even certain closed-form solutions can be derived what, naturally, makes the discussion of results particularly simple. In spite of the comparative simplicity of the model, the results are found to differ considerably from the classical ones, and in addition to quantitative, certain qualitative differences with respect to the classical solutions are established.

The aim of this series of lectures is to demonstrate how certain classical solutions of the elasticity theory can be "translated" into the slightly more general language of the couple-stress theory and to stress the fundamental differences between the classical and new results, thus increasing the possibility of determining the class of physical phenomena which can be explained only on the basis of the generalized model of elastic bodies, and which could not be explained as long as the simple Hookean elastic solid was considered.

2. Fundamental Equations

The derivation of the fundamental equations of the couple-stress theory presented in this section is mainly based on paper [1] by W. T. Koiter with certain modifications. The indicial notation (Cartesian tensors) is used throughout the paoer. The summation convention holds for repeated indices (except for the index n denoting the direction normal to the surface). The permutation symbol is denoted ϵ_{ijk} and assumes the values of +1, -1, 0, depending on whether the permutation of indices is even, odd or some indices are repeated. Vectors, denoted in print by bold-face letters, are underlined in this text.

Introduction Most of the classical text-books on elasticity and mechanics of deformable bodies do not introduce the notion of couple-stresses; it is tacitly assumed that the interaction of individual particles in the body can be reduced to simple forces or force vectors holding the entire body in equilibrium.

The simple elementary reasoning shows how the concept of couple-stresses can be introduced into the analysis of the problem of transmission of forces through continuous media.

Envisage an elementary parallelopiped frequently used in classical elasticity to derive the equilibrium conditions. The dimensions of the parallelopiped are usually assumed to be small enough to expect that the loads acting on a face, say, $x_1 =$ const., are uniformly distributed over the entire rectangle $dx_2 dx_3$; thus, the resultant force can be applied in the center of the rectangle which simplifies the further derivations (Fig. 1).

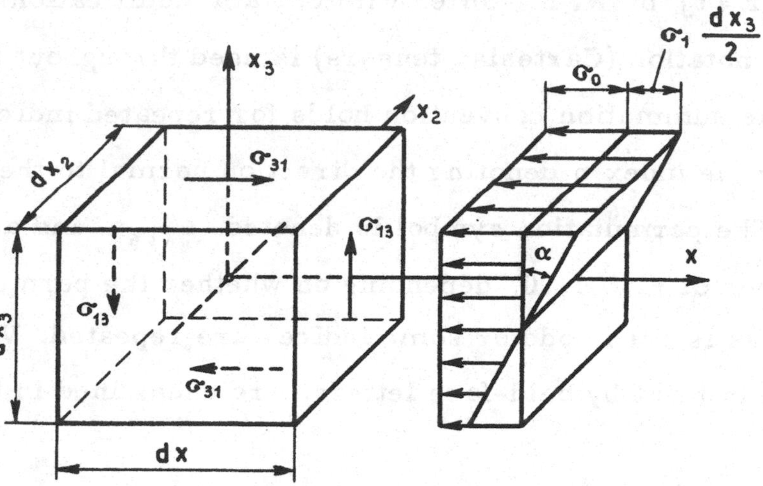

Fig. 1

The procedure is justified in the following manner.

Assume that the load acting on face $dx_2 dx_3$ consists of uniformly distributed shearing stresses σ_{13}, σ_{31} and of normal stresses $\sigma_{11} = \underset{0}{\sigma} + \underset{1}{\sigma} x_3$ being a linear function of x_3 and independent of x_2.

Writing down the conditions of equilibrium of moments of all forces acting on the element with respect to the x_2-axis, a contribution of the non-uniform σ_{11}-stress distribution (stresses acting on the opposite face are disregarded for sake of simplicity)

$$\frac{1}{2} \underset{1}{\sigma} \frac{dx_3}{2} dx_2 \frac{dx_3}{2} \frac{2}{3} dx_3 = \frac{1}{12} \underset{1}{\sigma} dx_2 (dx_3)^3$$

is added to the usual sum

$$(\sigma_{13} - \sigma_{31}) \, dx_1 \, dx_2 \, dx_3 \, .$$

Adding up these results one obtains

$$\left[(\sigma_{13} - \sigma_{31}) + \frac{1}{12} \frac{(dx_3)^2}{dx_1} \underset{1}{\sigma} \right] dx_1 \, dx_2 \, dx_3 \, .$$

Now, the expression in brackets contains terms of different order of magnitude: with $dx_i \rightarrow 0$ the last term involving the finite value $\underset{1}{\sigma} = \tan\alpha$ of the stress gradient and the first power of the infinitesimal magnitude dx_i can be neglected when compared with the first term in parenthesis and the usual result $\sigma_{13} = \sigma_{31}$ is obtained.

The above procedure cannot be, however, applied in the three following cases.

(a) The dimensions dx_i of the element cannot tend to zero owing to a finite size of the individual particles (grains) of the body; the body cannot be replaced by a continuous medium model. This case leads to the equations governing the behavior of bodies with microstructure where the stress tensor was found to be non-symmetric.

b) The coefficient σ_1 is not finite; if

$$\sigma_1 = \tan\alpha \rightarrow \infty ,$$

the stress gradient becomes infinite, and such situation is encountered at singular points of the stress field, at points of infinite stress concentration. The possibility of application of the couple-stress theory to the stress concentration problems was suggested by several authors : W. T. Koiter [1] E. Sternberg [2] and others.

c) The forces of interaction of the element with the surrounding body cannot be reduced to force vectors (represented by "arrows" in Fig. 1). If the forces of interaction are of a really polar nature (which, for instance, holds true in magnetic materials characterized by magnetic dipols-single magnetic poles do not appear in nature), the whole reasoning ceases to be of any physical meaning.

These remarks (which do not pretend to be of any rigorous mathematical value) and other, much more extensive studies of the problem of force transmission indicate that a detailed analysis of the so-called non-symmetric elasticity may prove interesting and may lead to results important both from the purely theoretical and the applied, technological point of view.

Equation of Continuity It is obvious that the derivation of the couple-stress theory has to be based on a number of physical assumptions identical with those used in the classical theory of elasticity. The fundamental laws of conservation of such quantities as the mass, energy, momentum and moment of momentum are considered as starting points of further considerations of the couple-stress theory.

Some of these laws,, however, are now expressed in a new form owing to the introduction of certain new variables and physical quantities into the modified theory. The well-known equation of continuity (conservation of mass) is based upon the simple notion of density ρ (mass per volume) and particle velocity $\underline{v}$, thus it remains completely unchanged and can be written in the usual integral form

$$\int_V \frac{\partial \rho}{\partial t}\, dV + \int_S \rho v_i n_i \, dS = 0$$

or in the local, differential form

$$\frac{\partial \rho}{\partial t} + \frac{\partial(\rho v_i)}{\partial x_i} = \frac{D\rho}{Dt} + \rho \frac{\partial v_i}{\partial x_i} = 0. \qquad (2.1)$$

Here S is the surface bounding the volume V, n_i are components of the unit normal to S, D/Dt is the material time derivative and v_i - components of the particle velocity vector.

Force-Stresses and Couple-Stresses Let us assume that the action of the surroundings of the volume element upon the element V can be represented by a certain distribution of forces $\underline{P}$ and couples $\underline{Q}$ on the surface S of element V. Let us assume, moreover, that the distribution $\underline{P}(S)$ and $\underline{Q}(S)$ has the following property : consider a point A on surface S of V (Fig. 2) belonging to an infinitesimal element dS of area ΔS.

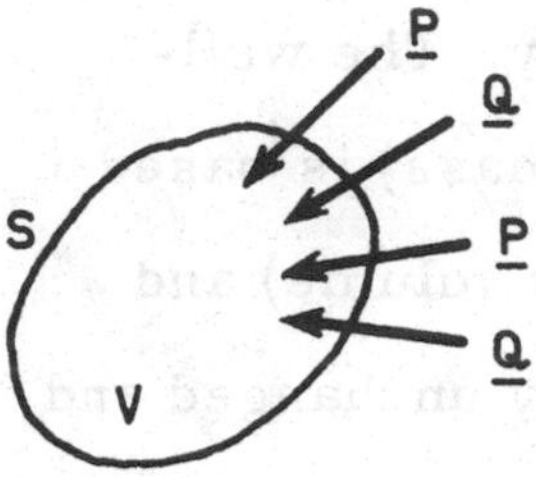

Fig. 2

Calculate the ratios

$$\frac{\Delta \underline{P}}{\Delta S} \quad \text{and} \quad \frac{\Delta \underline{Q}}{\Delta S}$$

If, for $\Delta S \to 0$ the ratios tend to definite limits $d\underline{P}/dS$ and $d\underline{Q}/dS$ denoted

$$\lim_{\Delta S \to 0} \frac{\Delta \underline{P}}{\Delta S} = \frac{d\underline{P}}{dS} = \overset{n}{\underline{p}},$$

$$\lim_{\Delta S \to 0} \frac{\Delta \underline{Q}}{\Delta S} = \frac{d\underline{Q}}{dS} = \overset{n}{\underline{q}},$$

these values are called contact stresses : $\overset{n}{\underline{p}}$ is the usual force-stress vector (or simply stress-vector) known from the classical elasticity and $\overset{n}{\underline{q}}$ is the couple-stress vector. Both $\underline{p}$ and $\underline{q}$ are, evidently, functions of position on surface S or, which is equivalent, functions of position in the body and of the direction normal to surface S what is manifested by the superscript n ($\underline{n}$ being the unit normal to S at point A).

The contact force and couple intensities, or force- and couple-stresses $\overset{n}{\underline{p}}$, $\overset{n}{\underline{q}}$ are vectors and they can be written in form of their components. In a Cartesian rectangular coordinate system x_i $(i=1,2,3)$ their components are denoted $\overset{n}{p}_i$ and $\overset{n}{q}_i$. Let us consider an infinitesimal tetrahedron (Fig. 3) with three faces normal to the axes x_1, x_2, x_3, respectively.

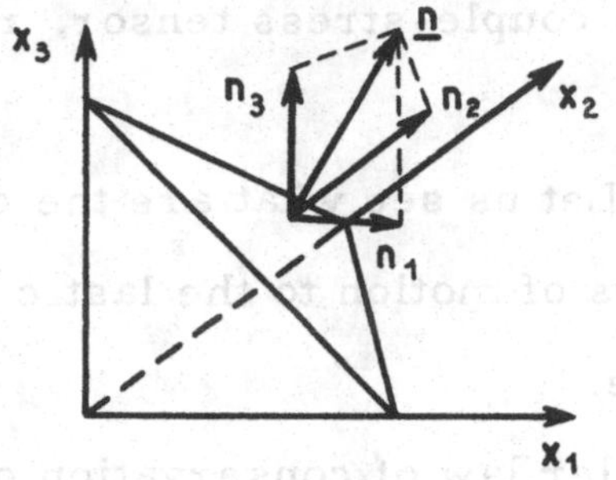

Fig. 3

Following the usual way of reasoning, it can easily be found that

$$\overset{n}{p}_1 = \overset{1}{p}_1 n_1 + \overset{2}{p}_1 n_2 + \overset{3}{p}_1 n_3$$

or, generally,

$$\overset{n}{p}_i = \sum_j \overset{j}{p}_i n_j . \tag{2.2}$$

Similarly,

$$\overset{n}{q}_i = \sum \overset{j}{q}_i n_j . \tag{2.3}$$

From the well-known quotient rule of tensor calculus and the fact that both $\overset{n}{p}_i$ and n_j are vector components, it follows that $\overset{j}{p}_i$ form the components of a tensor of second rank and can be, in Cartesian coordinates, written as σ_{ji}

and μ_{ji}. Thus

$$\overset{n}{p}_i = \sigma_{ji} n_j , \qquad \overset{n}{q}_i = \mu_{ji} n_j . \tag{2.4}$$

σ_{ji} and μ_{ji} are called the components of the force-stress tensor and the couple-stress tensor, respectively.

Conservation Laws Let us see what are the consequences of application of the laws of motion to the lastic medium of properties derived above.

The first Euler law of conservation of linear momentum states that

$$\frac{D}{Dt} \underline{\mathcal{P}} = \underline{\mathfrak{F}} . \tag{2.5}$$

Here D/Dt is the material time-derivative of momentum $\mathcal{P}$ of a definite portion $V(t)$ of the body at instant t; $\mathfrak{F}$ is the total force vector applied to the body.

The second law of conservation of the moment of momentum states that

$$\frac{D}{Dt} \underline{\mathcal{H}} = \underline{\mathcal{L}} , \tag{2.6}$$

where $\mathcal{H}$ denotes the total moment of momentum of all particles contained in $V(t)$, $\underline{\mathcal{L}}$ -the total torque applied to the element. The momentum and moment of momentum are expressed in terms of particle velocity vector $\underline{v}$ and density ϱ in the well-known manner

$$\underline{\mathfrak{P}} = \int\limits_{V(t)} \underline{v}\, \rho\, dV \,, \qquad \underline{\mathfrak{H}} = \int\limits_{V(t)} \underline{r} \times \underline{v}\, \rho\, dV.$$

Here $\underline{r}$ is the position vector of the particle under consideration.

The right-hand terms of Euler's laws (2. 5) and (2. 6) are calculated as follows. The force $d\underline{\mathfrak{F}}$ acting on element dV consists of the body force contribution equal to $\rho \underline{X}\, dV$ and the contact force contribution $\overset{n}{p}_i\, dS$ integrated over the surface of the element

$$\mathfrak{F}_i = \int\limits_V \rho X_i\, dV + \int\limits_S \overset{n}{p}_i\, dS \,.$$

The torque applied to a body element consists of four parts : The contribution of body couples $\rho \underline{Y}\, dV$ applied to the volume and moments produced by body forces $\underline{r} \times \rho \underline{X}\, dV$; the contact couple-stress vectors $\underline{q}$ and force-stress vectors $\underline{p}$ produce torques $(\overset{n}{\underline{q}} + \underline{r} \times \overset{n}{\underline{p}})\, dS$. Making use of the permutation symbol ϵ_{ijk} to express the vector-product components, the total torque is written (in Cartesian coordinates) with the aid of Eqs. (2. 2), (2. 3)

$$\mathfrak{L}_i = \int\limits_V \rho Y_i\, dV + \int\limits_V \rho\, \epsilon_{ijk}\, x_j X_k\, dV +$$

$$+\int_S \mu_{ji}\, n_j\, dS + \int_S \epsilon_{ijk}\, x_j \sigma_{\ell k} n_\ell\, dS.$$

Hence, the integral form of Euler's conservation laws is

$$\frac{D}{Dt}\int_{V(t)} \rho v_i\, dV = \int_V \rho X_i\, dV + \int_S \sigma_{ji} n_j\, dS, \tag{2.7}$$

$$\frac{D}{Dt}\int_{V(t)} \rho\, \epsilon_{ijk}\, x_j v_k\, dV = \tag{2.8}$$

$$= \int_V (\rho Y_i + \rho \epsilon_{ijk} x_j X_k)\, dV + \int_S (\mu_{ji} n_j + \epsilon_{ijk} x_j \sigma_{\ell k} n_\ell)\, dS.$$

These integral laws can be transformed into a more useful and simpler local form under certain assumptions concerning the differentiability of the stress distribution functions. Making use of the Gauss theorem

$$\int_S f\, n_i\, dS = \int_V f_{,i}\, dV,$$

where $f_{,i}$ denotes the partial derivative of f with respect to x_i, Eq. (2.7) is transformed into

$$\frac{D}{Dt}\int_{V(t)} \rho v_i\, dV = \int_{V(t)} (\rho X_i + \sigma_{ji,j})\, dV. \tag{2.9}$$

The material time-derivative appearing in (2.9) is transformed according to the rule

$$\frac{D}{Dt}\int_{V(t)} A\, dV = \int_V \left[\frac{\partial A}{\partial t} + \frac{\partial}{\partial x_j}(A v_j)\right] dV \tag{2.10}$$

and we are led to

$$\frac{D}{Dt}\int_{V(t)} \rho v_i\, dV = \int_V \left[\frac{\partial(\rho v_i)}{\partial t} + \frac{\partial}{\partial x_j}(\rho v_i v_j)\right] dV =$$

$$= \int_V \left\{ v_i \left[\frac{\partial \rho}{\partial t} + \frac{\partial}{\partial x_j}(\rho v_j)\right] + \rho \left[\frac{\partial v_i}{\partial t} + v_j \frac{\partial v_i}{\partial x_j}\right]\right\} dV. \tag{2.11}$$

The first expression in brackets of (2.11) vanishes owing to the continuity equation (2.1). The second bracket contains the material time derivative of $\underline{v}$, that is the acceleration. Substituting these results into (2.11) and (2.9) we are led to the condition

$$\int_V \left(\rho \frac{Dv}{Dt} - \rho X_i - \sigma_{ji,j}\right) dV = 0. \tag{2.12}$$

Since (2.11) holds for every portion V of the body, the integrand in Eq. (2.11) has to vanish and thus the final differential (local) form of the first Euler's law of motion is

$$\sigma_{ji,j} + \rho X_i = \rho \frac{Dv_i}{Dt}. \tag{2.13}$$

It is seen that this law is identical with the equations of motion known from classical elasticity.

The second law (2.8) is transformed - under the restrictions mentioned before - in the following manner. The right-hand side of (2. 8), with the aid of Gauss theorem, is written as

$$\int_V \left[\varrho Y_i + \mu_{ji,j} + \epsilon_{ijk}(\varrho x_j X_k + x_j \sigma_{\ell k,\ell} + \sigma_{jk})\right] dV$$

or, on using condition (2. 13), as

$$(2.14) \qquad \int_V \left[\varrho Y_i + \mu_{ji,j} + \epsilon_{ijk}\left(x_j \varrho \frac{Dv_k}{Dt} + \sigma_{jk}\right)\right] dV.$$

The left-hand term in (2. 8) is written, according to (2.10),

$$\epsilon_{ijk} \int_V \left[\frac{\partial(\varrho x_j v_k)}{\partial t} + \frac{\partial}{\partial x_\ell}(\varrho x_j v_k v_\ell)\right] dV =$$

$$= \epsilon_{ijk} \int_V \left[x_j \frac{\partial(\varrho v_k)}{\partial t} + x_j \frac{\partial(\varrho v_k v_\ell)}{\partial x_\ell} + \varrho v_k v_j\right] dV.$$

The product $\epsilon_{ijk} v_k v_j$ evidently vanishes; Eq. (2. 8) takes now the form

$$\int_V \left\{\varrho Y_i + \mu_{ji,j} + \epsilon_{ijk}\left[x_j \varrho \frac{Dv_k}{Dt} - \frac{\partial(\varrho v_k)}{\partial t} - \frac{\partial(\varrho v_k v_\ell)}{\partial x_\ell}\right] + \right.$$

$$\left. + \sigma_{jk}\epsilon_{ijk}\right\} dV = \int_V \left\{\varrho Y_i + \mu_{ji,j} + \epsilon_{ijk}\sigma_{jk} + \right.$$

$$\left. + \epsilon_{ijk} x_j \left[\varrho \frac{Dv_k}{Dt} - v_k\left(\frac{\partial \varrho}{\partial t} + \frac{\partial(\varrho v_\ell)}{\partial x_\ell}\right) - \varrho\left(\frac{\partial v_k}{\partial t} + v_\ell \frac{\partial v_k}{\partial x_\ell}\right)\right]\right\} dV.$$

The local, differential form of the second Euler law is then

$$\mu_{ji,j} + \epsilon_{ijk}\,\sigma_{jk} + \varrho Y_i = 0 \qquad (2.15)$$

In a static case Eqs. (2.13), (2.15) reduce to

$$\sigma_{ji,j} + \varrho X_i = 0 \qquad (2.16)$$

$$\mu_{ji,j} + \varrho Y_i + \epsilon_{ijk}\sigma_{jk} = 0\,. \qquad (2.17)$$

It is seen here that instead of the usual stress tensor symmetry required by the second law of motion, an additional set of differential equations (2.17) is obtained and no reason for σ_{ij} to be symmetric can be found. The state of stress in the body considered here is thus characterized by 9 components of the force-stress tensor σ_{ij} and 9 components μ_{ij} of the couple-stress tensor.

In order to further simplify the equations, the stress tensors σ_{ij} and μ_{ij} can be written as

$$\sigma_{ij} = \sigma_{(ij)} + \sigma_{[ij]} = s_{ij} + \overset{a}{\sigma}_{ij}\,,$$

$$\mu_{ij} = \mu\delta_{ij} + (\mu_{ij} - \mu\delta_{ij}) = \mu\delta_{ij} + m_{ij}\,,$$

where $s_{ij}=\frac{1}{2}(\sigma_{ij}+\sigma_{ji})$, $\overset{a}{\sigma}_{ij}=\frac{1}{2}(\sigma_{ij}-\sigma_{ji})$ are the symmetric and antisymmetric parts of the σ_{ij}-tensor, $\mu=\frac{1}{3}\mu_{ii}$, $m_{ij}=\mu_{ij}-\mu\delta_{ij}$ -the spherical and deviatoric parts of the μ_{ij}-tensor.

Eq. (2.17) can serve to express $\overset{a}{\sigma}_{ij}$ in terms of μ_{ij}, namely, multiplication of (2.17) by ϵ_{ipq} yields

$$\overset{a}{\sigma}_{pq}=-\frac{1}{2}\epsilon_{ipq}\left(m_{ji,j}+\mu_{,i}+\rho Y_i\right). \tag{2.18}$$

Eq. (2.16) is written in the form

$$s_{ji,j}+\overset{a}{\sigma}_{ji,j}+\rho X_i=0$$

or

$$s_{ji,j}-\frac{1}{2}\epsilon_{kji}\left[m_{\ell k,\ell j}+(\rho Y_k)_{,j}\right]+\rho X_i=0. \tag{2.19}$$

Eqs. (2.19) constitute a system of three second order differential equations of equilibrium replacing the six equations (2.16), (2.17). The symmetric part of σ_{ij} and the deviatoric part of μ_{ij} only enter these equations explicitly.

Strains In order to determine the appropriate set of strain components which, when added to the set of stresses, would form a complete system of variables describing the behavior of the body, let us calculate the rate at which work is done by external forces applied to volume element V of the body. The external loading consists of the body forces and couples $\underline{X},\underline{Y}$

and the surface forces and couples, $\underline{p}$ and $\underline{q}$. Multiplying these loads by the corresponding generalized displacement rates, the specific power (per unit mass) of external forces being denoted by $\dot{A}$, we are led to

$$\int_V \varrho \dot{A}\, dV = \int_V (\underline{X}\,\dot{\underline{u}} + \underline{Y}\,\dot{\underline{\omega}})\,\varrho\, dV + \int_S (\underline{p}\,\dot{\underline{u}} + \underline{q}\,\dot{\underline{\omega}})\, dS .$$

With the aid of Gauss formula

$$\int_V \varrho \dot{A}\, dV = \left[\varrho X_k \dot{u}_k + \varrho Y_k \dot{\omega}_k + (\sigma_{\ell k}\dot{u}_k + \mu_{\ell k}\dot{\omega}_k)_{,\ell}\right] dV. \qquad (2.20)$$

Here

$$\omega_k = \frac{1}{2}\,\epsilon_{kpq}\, u_{q,p} \qquad (2.21)$$

is the infinitesimal rotation vector.

Regrouping the terms in the integrand of (2. 20)

$$\int_V \varrho \dot{A}\, dV = \int_V \Big[(\varrho X_k + \sigma_{\ell k,\ell})\,\dot{u}_k + (\varrho Y_k + \mu_{\ell k,\ell})\,\dot{\omega}_k + $$
$$+ \sigma_{\ell k}\dot{u}_{k,\ell} + \mu_{\ell k}\dot{\omega}_{k,\ell}\Big] dV ,$$

and making use of Eqs. (2. 16), (2. 17) we obtain

$$\int_V \varrho \dot{A}\, dV = \int_V (-\epsilon_{kpq}\sigma_{pq}\dot{\omega}_k + \sigma_{\ell k}\dot{u}_{k,\ell} + \mu_{\ell k}\dot{\omega}_{k,\ell})\, dV .$$

Owing to relations

$$-\epsilon_{kpq}\sigma_{pq}\dot{\omega}_k = \overset{a}{\sigma}_{nm}\dot{u}_{n,m},$$

$$\sigma_{\ell k}\dot{u}_{k,\ell} = s_{\ell k}\dot{u}_{k,\ell} - \overset{a}{\sigma}_{k\ell}\dot{u}_{k,\ell},$$

$$\mu_{\ell k}\dot{\omega}_{k,\ell} = m_{\ell k}\dot{\omega}_{k,\ell}$$

the final form of the power of external forces is

$$\int_V \rho\dot{A}\,dV = \int_V (s_{\ell k}\dot{u}_{k,\ell} + m_{\ell k}\dot{\omega}_{k,\ell})\,dV. \tag{2.22}$$

It is seen here that no work is done by the skew-symmetric part of the force-stress tensor and by the spherical part of the couple-stress tensor.

Eq. (2.22) holds for any portion of the body ; tensor $s_{\ell k}$ being symmetric, only the symmetric part of $\dot{u}_{k,\ell}$ is essential; denoting, as usually, the infinitesimal strain tensor by

$$u_{(i,j)} = \frac{1}{2}(u_{i,j} + u_{j,i}) = \epsilon_{ij}$$

and, in addition, the infinitesimal torsion-flexure tensor by

$$\omega_{i,j} = \frac{1}{2}\epsilon_{ipq}u_{q,pj} = \varkappa_{ij},$$

Eq. (2.22) is reduced to the equation

$$\rho\dot{A} = s_{\ell k}\dot{\varepsilon}_{k\ell} + m_{\ell k}\dot{\varkappa}_{k\ell}. \tag{2.23}$$

Force-stress tensor σ_{ij}, couple-stress tensor μ_{ij}, strain tensor ε_{ij} and torsion-flexure tensor $\varkappa_{ij}$ form a full set of quantities describing the static behavior of the continuous medium under consideration.

Constitutive Equations The expression for the power of external loads (2.23) in an elastic body makes it possible to choose the suitable form of the strain energy function W . It is assumed here that the work done by external loads is entirely stored within the body in the form of elastic energy.

$$W = W(\varepsilon_{ij}, \varkappa_{ij}) .$$

If the elastic body is assumed to obey linear stress-strain relations and if the natural state of the body exists and is characterized by vanishing strains and stresses, then $W(\varepsilon_{ij}, \varkappa_{ij})$ should be an homogeneous quadratic function of strain components

$$W(\varepsilon_{ij}, \varkappa_{ij}) = \frac{1}{2} A_{ijkl}\, \varepsilon_{ij}\, \varepsilon_{kl} + B_{ijkl}\, \varepsilon_{ij}\, \varkappa_{kl} + \frac{1}{2} C_{ijkl}\, \varkappa_{ij}\, \varkappa_{kl} \tag{2.24}$$

The stress-strain relations derived from formula (2.24) have the form

$$\sigma_{ij} = \frac{\partial W}{\partial \varepsilon_{ij}} = A_{ijkl}\, \varepsilon_{kl} + B_{ijkl}\, \varkappa_{kl} , \tag{2.25a}$$

(2.25b) $$m_{ji} = \frac{\partial W}{\partial \varkappa_{ji}} = B_{klij}\,\varepsilon_{kl} + C_{ijkl}\,\varkappa_{kl}\,.$$

The total number of coefficients in (2.24) equals 3.81 = 243; the number of variables is : 6 independent force-stress components and 8 independent couple-stress components.

Owing to obvious symmetry requirements

$$A_{ijkl} = A_{klij} = A_{lkij} = A_{lkji} \qquad \text{(21 coefficients)},$$

$$C_{ijkl} = C_{klij} \qquad \text{(36 coefficients)},$$

$$B_{ijkl} = B_{jikl} \qquad \text{(48 coefficients)},$$

the number of independent coefficients is reduced to 105. This is true in the case of a body with no definite elastic symmetry properties. The aim of this derivation is to establish the possibly simplest model of a body able to transmit the couple-stresses. Hence, the full elastic symmetry (isotropy) is assumed here what confines the general form of the strain-energy function to terms involving the quadratic invariants of the strain components, i.e.

$$\varepsilon_{jj}\,\varepsilon_{ii}\,,\quad \varepsilon_{ij}\,\varepsilon_{ij}\,,\quad \varepsilon_{ij}\,\varkappa_{ij}\,,\quad \varkappa_{ij}\,\varkappa_{ij}\,,\quad \varkappa_{ij}\,\varkappa_{ji}$$

which leads to the following form of the stress-strain relations

$$s_{ij} = a\,\varepsilon_{kk}\,\delta_{ij} + b\,\varepsilon_{ij} + c\,\varkappa_{ij}\,,$$

$$m_{ij} = c\varepsilon_{ij} + d\varkappa_{ji} + e\varkappa_{ij} .$$

Additional requirements imply that a body subject to uniform, homogeneous deformation $\varepsilon_{ij} = \text{const.}$ resulting from linear displacements $u_i = a_{ij}x_j$ (thus, $\omega_{i,j} = \varkappa_{ij} \equiv 0$) no couple stresses appear: the body behaves like a classical Hookean elastic solid. This leads to the strain energy function

$$W = a\varepsilon_{ii}\varepsilon_{jj} + b\varepsilon_{ij}\varepsilon_{ij} + c\varkappa_{ij}\varkappa_{ij} + d\varkappa_{ij}\varkappa_{ji} \tag{2.26}$$

It is seen here that, in addition to the usual two elastic constants a, b, two additional coefficients c, d appear. Changing the notation and introducing the common notions of the shear modulus G, Young's modulus E and Poisson's ratio ν, Eq. (2.26) is written as

$$W = G\left[\frac{\nu}{1-2\nu}(\varepsilon_{ii})^2 + \varepsilon_{ij}\varepsilon_{ij} + 2\ell^2(\varkappa_{ij}\varkappa_{ij} + \eta\varkappa_{ij}\varkappa_{ji})\right]. \tag{2.27}$$

From the condition of positive definiteness of W in the neighbourhood of the neutral state it follows that $E > 0$, $0{,}5 > \nu > -1$ (as in the classical case), and that

$$\ell^2 \geqslant 0 , \quad -1 < \eta < +1 .$$

The notation of the additional elastic constant, ℓ^2, is justified by the fact that $\dim(\ell^2) = (\text{length})^2$, since

$$\dim(\ell^2) = \frac{\dim(\varepsilon_{ij})^2}{\dim(\varkappa_{ij})^2} = \text{length}^2$$

For the sake of simplicity it is sometimes assumed that the other, dimensionless constant $\eta = 0$, though, in general, this assumption is not justified.

Boundary Conditions The complete set of generalized forces and displacements which appears in the expression (2.20) for the power of external loads consists of six generalized force components p_i, q_i and six components of the generalized displacement vectors u_i, ω_i (i.e. the vectors of rotation and displacement). The natural assumption would thus be to expect a boundary value problem to be completely formulated by prescribing either the six geometric quantities or the six dynamic (force) quantities (leaving aside the mixed boundary value problem where a certain combinations of these quantities could be prescribed). It is evident, however, that six rotation and displacement components can not be independently prescribed on a surface bounding the body: if the outward normal $\underline{n}$ to surface S coincides, e.g., with the x_3-direction, the ω_3-component is completely determined by $u_1(x_1,x_2)$ and $u_2(x_1,x_2)$, since

$$\omega_3 = \frac{1}{2}\left(\frac{\partial u_2}{\partial x_1} - \frac{\partial u_1}{\partial x_2}\right). \tag{2.28}$$

The similar observation applied to dynamic boundary conditions which is easily demonstrated by the following argumentation. The six dynamic conditions are written as

$$p_i = \sigma_{ji} n_j = (s_{ji} + \overset{a}{\sigma}_{ji}) n_j, \tag{2.29}$$

$$q_i = \mu_{ji} n_j' = m_{ji} n_j + \mu n_i. \tag{2.30}$$

But, substituting here $\overset{a}{\sigma}_{ji}$ from Eq. (2.18)

$$p_i = \left[s_{ji} + \frac{1}{2} \epsilon_{jli} (m_{pl,p} + \mu_{,l} + \rho Y_l) \right] n_j. \tag{2.31}$$

From Eq. (2.30) the μ-invariant can be calculated

$$\begin{gathered} \mu n_i = q_i - m_{ji} n_j, \\ \mu = q_i n_i - m_{ji} n_j n_i = \overset{n}{q} - \overset{n}{m}. \end{gathered} \tag{2.32}$$

The first right-hand term represents the component of vector $\underline{q}$ normal to the surface; the second term, the normal component of m_{ij} acting on the surface element perpendicular to $\underline{n}$ (e.g. m_{33} in the case of $\underline{n}$ parallel to the x_3-axis).

Substituting (2.32) into (2.31) we are led to

$$q_i - m_{ji} n_j = (\overset{n}{q} - \overset{n}{m}) n_i \tag{2.33}$$

$$p_i = \left[s_{ji} + \frac{1}{2} \epsilon_{jli} (m_{pl,p} + \overset{n}{q}_{,l} - \overset{n}{m}_{,l} + \rho Y_l) \right] n_j. \tag{2.34}$$

Simple regrouping in (2.33), (2.34) yields

$$q_i - \overset{n}{q} n_i = m_{ji} n_j - \overset{n}{m} n_i \tag{2.35}$$

$$p_i - \frac{1}{2} \epsilon_{jli} \overset{n}{q}_{,l} n_j = s_{ji} + \frac{1}{2} \epsilon_{jli} (m_{pl,p} - \overset{n}{m}_{,l} + \rho Y_l) n_j. \tag{2.36}$$

Only the symmetric part of σ_{ji} and the deviatoric part of μ_{ji} appear in these conditions. The skew-symmetric part $\overset{a}{\sigma}_{ji}$ is expressed by relations (2.18) whereas the invariant μ remains indeterminate. Eqs. (2.35), (2.36) represent the required five boundary conditions expressed in stresses. It proves convenient to introduce the notions of reduced surface forces

$$\bar{p}_i = p_i - \frac{1}{2}\epsilon_{j\ell i}\overset{n}{q}_{,\ell}\, n_j \tag{2.37}$$

and reduced surface couples

$$\bar{q}_i = q_i - \overset{n}{q}\, n_i\,. \tag{2.38}$$

Owing to the fact that

$$\bar{\underline{q}}\,\underline{n} = \bar{q}_i n_i = q_i n_i - q_j n_j n_i n_i = 0\,,$$

vector $\bar{\underline{q}}$ is found to lie in a plane normal to $\underline{n}$ (tangent to S) and hence to possess two independent components only. Vector $\bar{\underline{p}}$ has three components: the scalar product $\bar{\underline{p}}\cdot\underline{n} = \underline{p}\cdot\underline{n}$, the normal component of $\underline{p}$ remains unchanged when compared to $\underline{p}$. The two tangential components of $\bar{\underline{p}}$ and $\underline{p}$ differ.

There remains to consider the singular lines on the surface S bounding the body, i.e. the lines dividing smooth portions of surfaces. Let us assume that both surfaces lying on two sides of the sharp edge are loaded by torques $\overset{n}{q}{}^{+}$, $\overset{n}{q}{}^{-}$ (Fig. 4). Applying here the elementary procedure known from the theory of plates (edge forces in simply supported polygonal plates), couples q^{+}, q^{-} are replaced by linear loads t^{+}, t^{-}

distributed uniformly along circles of radii r. The intensity of t is easily calculated from the condition

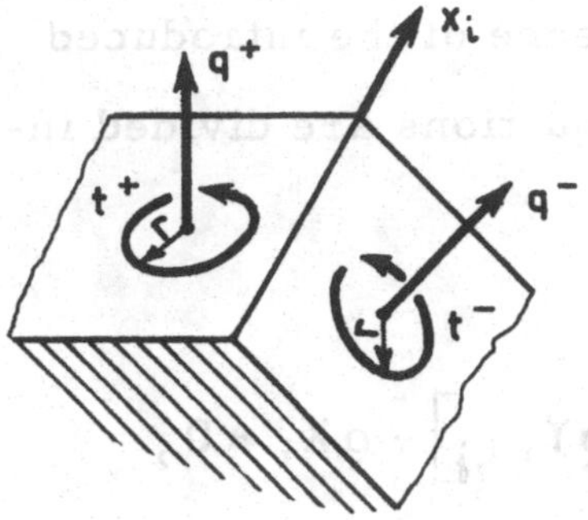

Fig. 4

$$\frac{t^{\pm} \cdot 2\pi r \cdot r}{\pi r^2} = q^{\pm}, \quad t^{\pm} = \frac{1}{2} q^{\pm}.$$

The horizontal resultant of these forces is evidently $2rt^+$ on the positive, and $-2rt^-$ on the negative face, which gives rise to a resultant line load of intensity

$$\bar{Q} = t^+ - t^- = \frac{1}{2}(\overset{n}{q}{}^+ - \overset{n}{q}{}^-)$$

acting along the edge. This result is confirmed by a rigorous consideration of the problem.

Thus, along a sharp edge of the surface bounding the body the internal couples μ_{ji} are set in equilibrium by a line load of intensity

$$\bar{Q} = \frac{1}{2}\left[\overset{n}{\mu}{}^+ - \overset{n}{\mu}{}^-\right] = \frac{1}{2}\left(\overset{n}{m}{}^+ - \overset{n}{m}{}^-\right).$$

$\overset{n}{\mu}$ is replaced by $\overset{n}{m}$ in this formula owing to the obvious assumption of continuity of the scalar μ (all tensor components of μ_{ij} are assumed to be continuous; the jump of $\overset{n}{\mu}$ results from the discontinuity of $\underline{n}$ across the edge).

Summary of Fundamental Equations The equations derived in this section governing the behavior of micropolar bodies are written together to demonstrate the influence of the introduced new element; the couple stresses. The equations are divided into the following groups.

a) Equations of equilibrium

$$s_{ji,j} - \frac{1}{2} \epsilon_{kji} \left[m_{\ell k,\ell j} + (\varrho Y_k)_{,j} \right] + \varrho X_i = 0 ;$$

b) Stress-strain relations

(2. 39)
$$s_{ij} = 2G \left(\varepsilon_{ij} + \frac{\nu}{1-2\nu} \varepsilon_{kk} \delta_{ij} \right),$$
$$m_{ij} = 4G\ell^2 \left(\varkappa_{ji} + \eta \varkappa_{ij} \right)$$

c) Geometric relations

$$\varepsilon_{ij} = u_{(i,j)} , \quad \varkappa_{ij} = \omega_{i,j}$$

with

(2. 40)
$$\omega_i = \frac{1}{2} \epsilon_{ipq} u_{q,p}$$

d) Boundary conditions

- expressed in displacements : vector $\underset{\sim}{u}$ and tangential components of $\underset{\sim}{\omega}$ are prescribed on the surface ;

- expressed in stresses :

(2. 41)
$$\left[s_{ji} + \frac{1}{2} \epsilon_{j\ell i} \left(m_{p\ell,p} - \overset{n}{m}_{,\ell} + \varrho Y_\ell \right) \right] n_j = \bar{p}_i ,$$
$$m_{ji} n_j - \overset{n}{m} n_i = \bar{q}_i ,$$

with reduced values of surface loads

$$\bar{p}_i = p_i - \frac{1}{2} \epsilon_{j\ell i} \overset{n}{q}_{,\ell} n_j ,$$

$$\bar{q}_i = q_i - \overset{n}{q} n_i .$$

Along a sharp edge of the surface

$$\frac{1}{2} \left(\overset{n}{m}^+ - \overset{n}{m}^- \right) = \bar{Q} ,$$

where $\bar{Q}$ - intensity of the line load. It should be mentioned here that also in the case of purely geometric boundary conditions the rotation vectors $\underline{\omega}^{\pm}$ cannot be arbitrarily prescribed on the adjoining surfaces; the proof is elementary and is based on the simple observation that the displacement on the sharp edge must be unique and independent of whether it is calculated from the rotation prescribed on the "positive" or "negative" surface.

It is to be mentioned here that when considering the phenomena occurring at sharp edges of a solid, the superiority of the couple-stress approach is readily manifested. In [3] Bogy and Sternberg show that the classical solution for an orthogonal elastic wedge, one edge of which is subjected to arbitrary shearing tractions, gives rise to certain singularities in the stress and rotation field. These singularities cannot be physically justified, although they obviously follow from the classical assumptions concerning the symmetry of the stress tensor. Within the frames of the couple-stress theory these singularities disappear.

Summing up, the number of unknowns and equations governing the couplc-stress theory is the following :

a) <u>General approach</u> : Unknowns : 3 displacements u_i, three rotations ω_i, 6 strains ε_{ij}, 9 torsion-flexure components $\varkappa_{ij}$, 9 force-stresses σ_{ij}, 9 couple-stresses μ_{ij} ; altogether 39 unknowns.

Equations : 6 eqs. of equilibrium, 14 stress-strain relations (2. 39), 18 geometric relations (2. 49); altogether 38 equations. The problem is "indeterminate", μ is the "redundant" value.

b) <u>Reduced approach</u> : The number of unknowns is reduced by four : μ and the antisymmetric components $\overset{a}{\sigma}_{ij}$. The number of equations is reduced by three (three eqs. 2. 19 instead of six eqs. 2. 16, 2. 17). Thus, the number of equations and unknowns is 35.

3. Generalized Kelvin's Solution

One of the most fundamental solutions in continuous medium mechanics is the solution of the problem of action of a concentrated force in an infinite, unbounded body. In classical elastic bodies the problem is known under the name of Kelvin's problem and the solutions is usually achieved with the aid of Love's stress function.

It has been earlier mentioned that the stress concentration problems may serve as a good starting point for possible verification of the couple stress theory, since then the classical assumptions concerning the mechanism of force transmission may be questioned. On the other hand, once the solution of the problem is found, the solution to the problem of arbitrary body force distribution within the elastic micropolar continuum is easily achieved by simple - at least formally simple - integration; it is the usual Green function procedure which can be applied as long as infinite regions are concerned. For finite boundaries the corresponding boundary effects are to be taken into consideration.

The solution to this problem has been presented in a paper by J.M. Doyle [4] in 1966. The solution to be presented here is based on Doyle's paper, though it slightly differs from the original derivation, being less rigorous and considerably shorter.

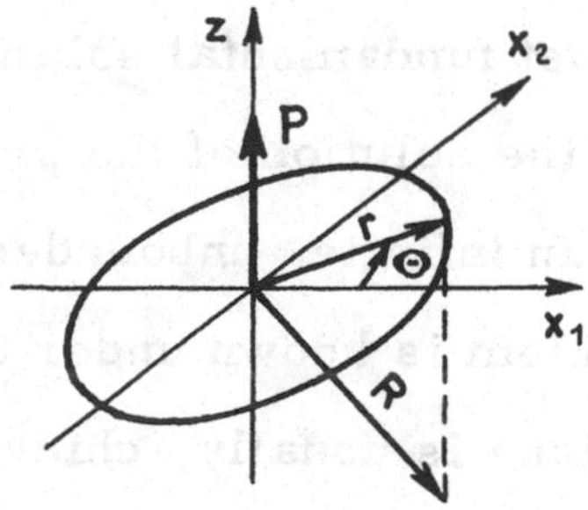

Fig. 5

Let us recall that the solution to the classical Kelvin's problem in cylindrical coordinates (r, Θ, z) reads as follows :

$$u_r = \frac{P}{4\pi G(1-\nu)} \frac{rz}{R^3},$$

$$u_\Theta = 0,$$

(3. 1) $$u_z = \frac{P}{16\pi G(1-\nu)} \left(\frac{3-4\nu}{R} + \frac{z^2}{R^3} \right),$$

$$R^2 = r^2 + z^2$$

It is seen from these formulae that at small distances from the origin 0 (where the concentrated load is applied) the displacements u_r , u_z possess a singularity of order one and increase indefinitely as $1/R$ with R tending to zero. With increasing R the displacements vanish as $1/R$; these properties of the displacements field are characteristic for the classical elastic model.

Let us return to the micropolar medium described by the equations of motion (2. 13) and (2. 15) of the preceding Section. These equations have been reduced to a simplified, reduced system of second order differential equations (2. 19) in the static case. Preserving the term due to inertia appearing in (2. 13) and absent from (2. 19) this equation is written as

$$\sigma_{ji,j} - \frac{1}{2}\epsilon_{kji}\left[m_{\ell k,\ell j} + \rho Y_{k,j}\right] + \rho X_i = \rho \ddot{u}_i . \tag{3. 2}$$

Here the corresponding assumptions concerning small displacements and velocities have been made (partial time derivative replacing the material ones, density ρ assumed constant).

Equation (3. 2) is now transformed to obtain the equations of motion in displacements, according to (2. 3)

$$\sigma_{ji} = G u_{i,j} + G u_{j,i} + \lambda u_{k,k}\delta_{ij} ,$$

$$m_{ij} = 2G\ell^2(\epsilon_{jpq}u_{q,p\ell\ell i} + \eta\epsilon_{\ell pq}u_{q,pk\ell j}) ,$$

and the second term in m_{ij} vanishes due to indices p, ℓ appearing twice in a symmetric and skew-symmetric tensors. Substituting this into (3. 2) we are led to the equation

$$\frac{G}{\rho}u_{i,jj} + \frac{\lambda + G}{\rho}u_{j,ji} + \frac{G}{\rho}\ell^2\epsilon_{ijk}\epsilon_{kpq}u_{q,pj\ell\ell} +$$

$$+ X_i + \frac{1}{2}\epsilon_{ijk}Y_{k,j} = \ddot{u}_i . \tag{3. 3}$$

Introducing now the new parameters c_L , c_T which

denote the velocities of longitudinal and transversal elastic waves, respectively,

$$c_L^2 = \frac{\lambda + 2G}{\varrho}, \quad c_T^2 = \frac{G}{\varrho}$$

and denoting by F_i the components of reduced body forces

$$\underline{F} = \underline{X} + \frac{1}{2}\,\mathrm{curl}\,\underline{Y},$$

$$F_i = X_i + \frac{1}{2}\epsilon_{ijk} Y_{k,j}$$

Eq. (3. 3) is rewritten in the form

$$\begin{aligned} &c_T^2 \nabla^2 u_i + (c_L^2 - c_T^2) u_{j,ji} + \\ &+ c_T^2 \ell^2 \epsilon_{imr} \epsilon_{rkj} \nabla^2 u_{j,km} + F_i = \ddot{u}_i . \end{aligned} \tag{3. 4}$$

It is now assumed that the motion of the medium is harmonic in time, i. e.

$$u_i(x_j, t) = u_i(x_j)\exp(i\omega t)$$

where ω-the angular frequency of vibration. Then, Eq. (3. 4) takes the form

$$\begin{aligned} &(c_L^2 - c_T^2) u_{j,ji} + c_T^2 \nabla^2 u_i + \\ &+ c_T^2 \ell^2 \epsilon_{imr} \epsilon_{rkj} \nabla^2 u_{j,km} + \omega^2 u_i = -F_i \end{aligned} \tag{3. 5}$$

These are the modified Navier equations of harmonic motion of a micropolar continuum; u_i denote here, in constrast to Eq. (3. 4), the amplitudes of vibration.

In order to integrate the system (3. 5), the generalized Galerkin vector is introduced. In classical elasticity the Galerkin-Papkovich vector G'_i is defined in the following manner

$$\underline{u} = \frac{1-\nu}{G} \nabla^2 \underline{G}' - \frac{1}{2G} \operatorname{grad} \operatorname{div} \underline{G}', \tag{3.6}$$

and inserted into the Navier equations proves to satisfy a simple nonhomogeneous biharmonic equation

$$\nabla^2 \nabla^2 \underline{G}' = -\underline{X}/(1-\nu)$$

In the case od Eqs. (3. 5) involving additional terms due to couple-stresses and inertia, formula (3. 6) has to be assumed in a slightly more complicated manner

$$\underline{u} = \nabla^2 \underline{G} + a \operatorname{grad} \operatorname{div} \underline{G} + b\,\underline{G} + c\,\nabla^2 \operatorname{grad} \operatorname{div} \underline{G} \tag{3.7}$$

with constants a, b, c to be found from the requirement that the governing equations for $\underline{G}$ should be as simple as possible. Inserting (3. 7) into (3. 5) we conclude that with

$$a = -\frac{c_L^2 - c_T^2}{c_L^2}, \quad b = \frac{\omega^2}{c_L^2}, \quad c = -\frac{c_T^2 \ell^2}{c_L^2}$$

Eq. (3.7) takes the form

$$u_i = \nabla^2 G_i - \frac{c_L^2 - c_T^2}{c_L^2} G_{j,ji} - \frac{c_T^2 \ell^2}{c_L^2} \nabla^2 G_{j,ji} + \frac{\omega^2}{c_L^2} G_i$$

(3.8)

and the Galerkin-Somigliana vector G_i satisfies the sixth order differential equation

$$\left(\nabla^2 + \frac{\omega^2}{c_L^2}\right)\left(\ell^2 \nabla^2 \nabla^2 - \nabla^2 - \frac{\omega^2}{c_T^2}\right) G_i = \frac{F_i}{c_T^2} . \qquad (3.9)$$

With $\omega \to 0$ and $\ell \to 0$, this reduces to the biharmonic equation mentioned before (the different coefficient being, of course, immaterial).

It can be proved that, similarly as in the classical elasticity, representation (3.8) is complete, that is : for an arbitrary field of displacements u_i it is always possible to find such functions G_i which generate the prescribed u_i field according to (3.8) (the practical effective determination of $\underline{G}$ in an explicit form is another problem).

To integrate Eq. (3.9) let us first observe that a particular integral of the equation

$$\nabla^2 \varphi_1 - k^2 \varphi_1 = f_1(x_i) \qquad (3.10)$$

can be written in the form [5]

$$\varphi_1(x_i) = -\frac{1}{4\pi}\int \frac{e^{-k\varrho} f(\xi_i)}{\varrho}\, dv(\xi_i) \qquad (3.11)$$

where $\varrho = \varrho(x,\xi) = \sqrt{(x_i-\xi_i)(x_i-\xi_i)}$ and the integration is extended over the entire region of definiteness of f_1. Substitution of (3.11) into (3.10) leads to the result

$$-(\nabla^2 - k^2)\frac{1}{4\pi}\int_D \frac{\exp[-k\varrho(\xi,x)]}{\varrho(\xi,x)} f_1(\xi)\, d\xi^3 =$$

$$= -(\nabla^2 - K^2)\frac{1}{4\pi}\left[\int_{D-S_\varepsilon} + \int_{S_\varepsilon}\right]$$

where S_ε denotes a small sphere of radius ε centered at the point $\xi_i = x_i$, i.e. $\varrho = 0$ (the singular point of the integrand). The first term vanishes owing to the simple identity

$$\nabla^2(e^{-k\varrho}\varrho^{-1}) = k^2 e^{-k\varrho}\varrho^{-1}$$

The second integral is transformed according to the Gauss theorem

$$-(\nabla^2 - k^2)\int_{S_\varepsilon} \frac{e^{-k\varrho(\xi,x)} f_1(\xi)}{\varrho(\xi,x)}\, d\xi^3 = -\int_{\sigma_\varepsilon} \frac{\partial}{\partial n}\left[\frac{e^{-k\varrho} f_1}{\varrho}\right] d\sigma_\varepsilon = 4\pi f_1(x_i)$$

which satisfies (3.10).

Equation (3.9) can be rewritten in the following manner

$$(\nabla^2 + \alpha^2)(\nabla^2 + \beta^2)(\nabla^2 - \gamma^2)\, G_i = \frac{F_i}{\ell^2 c_T^2}, \qquad (3.12)$$

where

$$\alpha^2 = \frac{\omega^2}{c_L^2}, \qquad \beta^2 = \frac{1}{2\ell^2}\left(\sqrt{\frac{4\omega^2\ell^2}{c_T^2}+1}-1\right)$$

$$\gamma^2 = \frac{1}{2\ell^2}\left(\sqrt{\frac{4\omega^2\ell^2}{c_T^2}+1}+1\right)$$

are positive numbers.

The solution of Eq. (3.12) is now written as a linear combination of solutions of the type (3.11)

(3.13) $$G_i = -\frac{1}{4\pi\ell^2 c_T^2}\int F_i(\xi)\left(A\,e^{i\alpha\rho}+Be^{i\beta\rho}+Ce^{-\gamma\rho}\right)\frac{d\xi^3}{\rho}$$

with

(3.14)
$$A = \frac{1}{(\alpha^2-\beta^2)(\alpha^2+\gamma^2)},$$

$$B = -\frac{1}{(\alpha^2-\beta^2)(\beta^2+\gamma^2)},$$

$$C = \frac{1}{(\alpha^2+\gamma^2)(\beta^2+\gamma^2)}.$$

<u>Remark</u>. Solution (3.13) may be easily obtained by the simple reasoning based on symbolic notation, D^2 replacing the Laplacean ∇^2. Eq. (3.12) is written as

(3.15) $$(D^2+\alpha^2)(D^2+\beta^2)(D^2-\gamma^2)\,G = \frac{F_i}{\ell^2 c_T^2}$$

and its solution is formally expressed as

$$G_i = \frac{1}{(D^2+\alpha^2)(D^2+\beta^2)(D^2-\gamma^2)} \frac{F_i}{\ell^2 c_T^2}$$

Decomposing the operator into "simple fractions"

$$G_i = \left[\frac{A}{D^2+\alpha^2} + \frac{B}{D^2+\beta^2} + \frac{C}{D^2-\gamma^2} \right] \frac{F_i}{\ell^2 c_T^2}$$

with A, B, C given by (3.14), the solution of (3.15) is found to consist of the three solutions of the (3.10)-type equations

$$(D^2+\alpha^2)\, G_i^{(1)} = \frac{A F_i}{\ell^2 c_T^2} ,$$

$$(D^2+\beta^2)\, G_i^{(2)} = \frac{B F_i}{\ell^2 c_T^2} , \qquad (3.16)$$

$$(D^2-\gamma^2)\, G_i^{(3)} = \frac{C F_i}{\ell^2 c_T^2} .$$

Combining (3.16) with the previously presented formula (3.11) ($i\alpha$ and $i\beta$ replace $-k$), the solution of the sixth order differential equation has the form (3.13).

The further derivation of the fundamental singularity presented in the paper by J.M. Doyle was based on the theorem of reciprocity of works done by external forces. Another method, less rigorous but comparatively simple, will be presented here. The results are identical.

Let us assume that the field of body forces appearing on the right-hand side of (3.5) consists of a single concentrated force P_j acting at the origin of the coordinate system. This is formally achieved by assuming first a uniform body force distribution within the volume $V = d^3$ of a cube centered at $x_i = 0$ and zero body forces without the cube; the resultant body force vector is

$$P_j = \varrho X_j d^3 .$$

With d tending to zero the body force intensity increases in order to keep the resultant force constant. In the limit the body force distribution is reduced to the triple Dirac delta distribution

$$X_j = \frac{P_j}{\varrho}\,\delta(x_1)\,\delta(x_2)\,\delta(x_3) = \frac{P_j}{\varrho}\,\delta^3(x_i).$$

Inserting this expression into formula (3.13) and making use of the fundamental δ-distribution property we are led to

$$(3.17)\qquad G_i = -\frac{P_j\,\delta_{ij}}{4\pi\ell^2 c_T^2\varrho}\left(A\,\frac{e^{i\alpha R}}{R} + B\,\frac{e^{i\beta R}}{R} + C\,\frac{e^{-\gamma R}}{R}\right)$$

which represents the Galerkin vector corresponding to the concentrated force P_j.

The singular displacement field generated by (3.17) is

obtained by substituting the expression for G_i into (3.8). All required differentiations performed, the result is written as

$$u_i = \frac{P_j}{4\pi \ell^2 \rho c_T^2}\left\{\frac{1}{c_L^2}\left[A\left(c_L^2 - c_T^2 - c_T^2\ell^2\alpha^2\right)\frac{e^{i\alpha R}}{R} + \right.\right.$$
$$+ B\left(c_L^2 - c_T^2 - c_T^2\ell^2\beta^2\right)\frac{e^{i\beta R}}{R} + C\left(c_L^2 - c_T^2 + c_T^2\ell^2\gamma^2\right)\left.\frac{e^{-\gamma R}}{R}\right]_{,ij} +$$
$$\left. + \left[B(\beta^2 - \alpha^2)\frac{e^{i\beta R}}{R} - C(\gamma^2 + \alpha^2)\frac{e^{-\gamma R}}{R}\right]\delta_{ij}\right\}. \qquad (3.18)$$

Here $R = R(x_i) = (x_i x_i)^{1/2}$.

From this general formula certain limiting cases can be,more or less easily, derived. For instance, in the static case with $\omega \to 0$ having in mind that for $\omega << 1$

$$\alpha = \frac{\omega}{c_L}, \quad \beta \cong \frac{\omega}{c_T}, \quad \gamma = \frac{1}{\ell} + \frac{\omega}{c_T}$$

the limiting procedure leads to

$$u_i = \frac{P_j}{4\pi\rho c_T^2}\left\{\left(\frac{1 - e^{-R/\ell}}{R}\right)\delta_{ij} + \right.$$
$$\left. + \left[-\frac{c_L^2 - c_T^2}{c_L^2}R + \ell^2\left(\frac{e^{-R/\ell} - 1}{R}\right)\right]_{,ij}\right\}. \qquad (3.19)$$

It is to be remembered here that, as usual in the case of complex potentials, the real part of the right-hand expression in (3.18) corresponds to actual displacements; the "Re"-

symbol is here omitted.

If, instead of ω, ℓ is assumed to tend to zero, the limiting procedure leads to the solution valid in classical elasticity under steady vibrations

$$(3.20)\qquad u_i = \frac{P_j}{4\pi\varrho\omega^2}\left\{\left[\frac{e^{i\alpha R}}{R} - \frac{e^{i\beta R}}{R}\right]_{,ij} + \frac{1}{\alpha^2}\,\frac{e^{i\alpha R}}{R}\,\delta_{ij}\right\}.$$

Finally, passing to the case of $\omega = \ell = 0$, i.e. to the classical Kelvin problem, we obtain

$$(3.21)\qquad u_i = \frac{P_j}{4\pi G}\left\{\left[-\frac{c_L^2 - c_T^2}{2c_L^2}\,R\right]_{,ij} + \frac{1}{R}\,\delta_{ij}\right\}$$

in accordance with the formula quoted at the beginning of this Section.

The formulae derived above can be subject to detailed discussion in order to point out the main differences between the classical and couple-stress solutions. Here let us point out that the order of singularity of solutions for u_i is not equal to three as it could be guessed at first sight from Eq. (3.18). It is easily verified that the term $[\;]_{,ij}$ in this equation can be transformed in such a way that the coefficients multiplying the functions $R^{-1}\cos\alpha R$, $R^{-1}\cos\beta R$, $R^{-1}\exp(-\gamma R)$

can be grouped as follows

$$\frac{c_L^2 - c_T^2}{c_L^2}\,(A + B + C)$$

and

$$\frac{c_L^2\,\ell^2}{c_L^2}\,(A\alpha^2 + B\beta^2 - C\gamma^2)$$

and, due to (3.14), these coefficients thus vanish decreasing the order of singularity. The same phenomenon is observed in Eq. (3.19) where the coefficients multiplying the term $\frac{1}{R}$ exponentially tend to zero as $R \to 0$ and decrease the order of singularity though leaving the possibility of second order singularities to remain. However, let us investigate this equation (corresponding to the static case, $\omega = 0$, and $\ell \neq 0$); assume the force $\underline{P}$ to act vertically : $P_1 = P_2 = 0$, $P_3 = P$; owing to relations

$$\frac{\partial^2}{\partial z^2}(R) = \frac{1}{R} - \frac{z^2}{R^3},$$

$$\frac{\partial^2}{\partial z^2}\left(\frac{1}{R}\right) = -\frac{1}{R^3} + \frac{3z^2}{R^5},$$

$$\frac{\partial^2}{\partial z^2}\left(\frac{e^{-R/\ell}}{R}\right) = \left(\frac{3z^2}{R^5} + \frac{3z^2}{\ell R^4} + \frac{z^2/\ell^2 - 1}{R^3} - \frac{1}{\ell R^2}\right)e^{-R/\ell},$$

the variation of u_3 along the vertical x_3-axis ($x_1 = x_2 = 0$, $x_3 = R$) is described by

$$u_3(0,0,z) =$$

$$= \frac{P}{4\pi G}\left\{\ell^2\left[\left(\frac{2}{z^3}+\frac{2}{\ell z^2}+\frac{1}{\ell^2 z}\right)e^{-z/\ell}-\frac{2}{z^3}\right]+\right.$$

$$\left. + \frac{1-e^{-z/\ell}}{z}\right\}. \tag{3.22}$$

For small values of R (small with respect to ℓ), expanding the exponential function into power series and preserving the first three terms of the expansion, all singular terms in (3.22) cancel out leaving us with the finite (!) value of the limit

$$\lim_{z\to 0} u_3(0,0,z) = \mathring{u}$$

namely

$$\mathring{u} = \frac{P}{6\pi G\ell}$$

This result was not mentioned in Doyle's paper and it certainly needs further discussion and explanation.

If, however, the point of application of the concentrated force is approached from a different direction, the solution remains singular, the order of singularity being one; namely

$$u_3 = \frac{P}{4\pi G}\left\{\frac{2}{3\ell} + \left(\frac{1}{R} - \frac{z^2}{R^3}\right)\left(\frac{1}{2} - \frac{c_L^2 - c_T^2}{c_L^2}\right)\right\} =$$

$$= \frac{P}{6\pi G\ell}\left[1 + \frac{3}{4}\,\frac{\nu}{1-\nu}\,\frac{\ell}{R}\left(1 - \frac{z^2}{R^2}\right)\right]. \qquad (3.23)$$

It is seen from the above formula that the coefficient multiplying the singularity $1/R$ is independent of ℓ,

$$K = \frac{P\nu}{8\pi G}\,\frac{1}{1-\nu}$$

but is different from the corresponding coefficient in the classical solution (3.1) in spite of the fact that the solution (3.19) as a whole transfroms with ℓ tending to 0 to the classical solution (3.21). A similar phenomenon will be encountered in our further considerations presented in Section 5.

4. Plane State of Strain

Among all theoretical problems of the theory of elasticity, the two-dimensional problems of plane strain (or stress) belong to the most frequently considered and discussed ones; this is caused, first of all, by the relative simplicity of the governing equations owing to the reduced number of unknown components of the displacement, stresses and deformation, and by a wider range of technical means of experimental verification of the theoretical results. In this Section we are going to derive the fundamental equations governing the behavior of micropolar elastic continua in a two-dimensional state of strain.

To this end let us turn to the case of a plane state of strain occurring in the plane perpendicular to the x_3-axis. It is assumed that the displacement vector $\underline{u}$ possesses only two non-vanishing components

$$\begin{aligned} &\underline{u} = \underline{u}\,(u_1, u_2), \quad u_3 = 0, \\ &u_1 = u_1(x_1, x_2), \quad u_2 = u_2(x_1, x_2). \end{aligned} \tag{4.1}$$

In view of this assumption the strain components are reduced in number to three independent components of the strain tensor

$$\varepsilon_{11} = u_{1,1}, \quad \varepsilon_{22} = u_{2,2}, \quad \varepsilon_{12} = \varepsilon_{21} = \frac{1}{2}(u_{1,2} + u_{2,1}), \tag{4.2}$$

and, in view of the only non-vanishing rotation vector

$$\omega_3 = \frac{1}{2}(u_{2,1} - u_{1,2}), \quad \omega_1 = \omega_2 = 0 \tag{4.3}$$

to two independent components of the torsion-flexure tensor

$$\varkappa_{31} = \omega_{3,1} = \frac{1}{2}(u_{2,11} - u_{1,12}),$$
$$\varkappa_{32} = \omega_{3,2} = \frac{1}{2}(u_{2,12} - u_{1,22}). \tag{4.4}$$

The remaining $\varkappa_{ij} = 0$.

In view of the above simplifications expression (2.27) for the strain energy function in an elastic, isotropic and centrosymmetric solid is explicitly written as

$$W = G\left[\frac{\nu}{1-2\nu}(\varepsilon_{11} + \varepsilon_{22})^2 + \varepsilon_{11}^2 + \varepsilon_{22}^2 + \varepsilon_{12}^2 + \right.$$
$$\left. + 2\ell^2(\varkappa_{31}^2 + \varkappa_{32}^2)\right], \tag{4.5}$$

the term multiplying the fourth, dimensionless constant η vanishing due to (4.4).

From Eq. (4.5) the stress-strain relations are derived, since, due to (2.25),

$$s_{ij} = \partial W / \partial \varepsilon_{ij}, \quad m_{ij} = \partial W / \partial \varkappa_{ji}$$

we obtain

$$s_{11} = \frac{2G\nu}{1-2\nu}(\varepsilon_{11} + \varepsilon_{22}) + 2G\varepsilon_{11}, \tag{4.6a}$$

$$s_{22} = \frac{2G\nu}{1-2\nu}(\varepsilon_{11} + \varepsilon_{22}) + 2G\varepsilon_{22},$$

(4.6b)
$$s_{11} + s_{22} = \frac{2G}{1-2\nu}(\varepsilon_{11} + \varepsilon_{22}),$$

$$s_{12} = 2G\varepsilon_{12}.$$

The remaining components of σ_{ij} do not contribute to the elastic potential and hence cannot be directly calculated from (4.5). Using relation (2.39) it is found that

(4.7)
$$s_{13} = s_{23} = 0,$$

$$s_{33} = \frac{2G\nu}{1-2\nu}(\varepsilon_{11} + \varepsilon_{22}) = \nu(s_{11} + s_{22}).$$

Finally, the deviatoric couple-stress components resulting from (4.5) are

(4.8)
$$m_{13} = 4G\ell^2\varkappa_{31},$$

$$m_{23} = 4G\ell^2\varkappa_{32},$$

$$m_{11} = m_{22} = m_{33} = m_{12} = m_{21} = 0.$$

The remaining two components m_{13}, m_{23} do not contribute to to the elastic energy (cf. (2.23)) and are again calculated directly from (2.39)

$$m_{31} = 4G\ell^2\eta\varkappa_{31}, \quad m_{32} = 4G\ell^2\eta\varkappa_{32}. \tag{4.9}$$

It is to be observed here that the notation for $\varkappa_{ij}$ accepted by W. T. Koiter [1] and in our considerations differs from the notations used in [2] where $\varkappa_{ij}$ is taken as $\omega_{j,i}$ and not $\omega_{i,j}$.

Passing to the equations of motion it is seen that Eq. (2.13), under small displacements and velocities, reduces to

$$\begin{aligned} &\sigma_{11,1} + \sigma_{21,2} + \overset{a}{\sigma}_{31,3} + \rho X_1 = \rho\ddot{u}_1, \\ &\sigma_{12,1} + \sigma_{22,2} + \overset{a}{\sigma}_{32,3} + \rho X_2 = \rho\ddot{u}_2, \\ &\overset{a}{\sigma}_{13,1} + \overset{a}{\sigma}_{23,2} = 0 \end{aligned} \tag{4.10}$$

where it has been assumed that the body and inertia forces in the x_3-direction are absent.

The other group of the equations of motion (2.17) under the assumption that no body couples act in the medium has the form

$$\begin{aligned} &\mu_{,1} + 2\overset{a}{\sigma}_{23} = 0, \\ &\mu_{,2} + 2\overset{a}{\sigma}_{31} = 0, \\ &\mu_{,3} + m_{13,1} + m_{23,2} + 2\overset{a}{\sigma}_{12} = 0. \end{aligned} \tag{4.11}$$

Here in Eqs. (4.10) and (4.11) the skew-symmetric σ_{ij} components and the invariant μ of the μ_{ij} - tensor appear; since they were not encountered in the previously presented constitutive equations and, hence, cannot be expressed in terms of displacements, no information concerning their possible independence of x_3 can be deduced. It will be later seen that this fact does not influence the final results since, as it was shown in Section 2, the mentioned components can entirely be eliminated in the course of further considerations. For the sake of simplicity, however, and conforming to the general idea of plane strain requiring the state of stress, strain and deformation to be independent of one coordinate, let us make the assumption that σ_{ij} and μ_{ij} are independent of x_3, and $\sigma_{13} = \sigma_{23} = \sigma_{31} = \sigma_{32} = 0$, hence simplifying the equations of motion to

$$\begin{aligned} &\sigma_{11,1} + \sigma_{21,2} + X_1 = \rho\ddot{u}_1, \\ (4.12) \qquad &\sigma_{12,1} + \sigma_{22,2} + X_2 = \rho\ddot{u}_2, \\ &m_{13,1} + m_{23,2} + \sigma_{12} - \sigma_{21} = 0, \end{aligned}$$

whereas the remaining equations (4.11) require only the indeterminate invariant μ to be constant throughout the body.

Proceeding as in Sect. 2, the skew-symmetric parts of σ_{12} can be eliminated from Eqs. (4.12) to lead to a simplified system of two equations ,

$$\sigma_{11,1} + \sigma_{12,2} + \frac{1}{2}(m_{13,1} + m_{23,2}) + X_1 = \rho\ddot{u}_1,$$
$$\sigma_{12,1} + \sigma_{22,2} - \frac{1}{2}(m_{13,1} + m_{23,2}) + X_2 = \rho\ddot{u}_2 \qquad (4.12a)$$

In the case when the problem is formulated and solved in stresses and strains (dynamic boundary conditions), the compatibility conditions are to be taken into consideration in order to make sure that the solutions do not violate the assumption of continuity of the medium after deformation. In the static case of plane strain and zero body forces these conditions quoted in [2] and derived in [6] are the following

$$\sigma_{11,22} + \sigma_{22,11} - \nu\nabla^2(\sigma_{11} + \sigma_{22}) = 2\sigma_{12,12},$$

$$m_{13,2} = m_{23,1},$$

$$m_{13} = -2\ell^2[\sigma_{11} - \nu(\sigma_{11} + \sigma_{22})]_{,2} + 2\ell^2\sigma_{12,1}, \qquad (4.13)$$

$$m_{23} = 2\ell^2[\sigma_{22} - \nu(\sigma_{11} + \sigma_{22})]_{,1} - 2\ell^2\sigma_{12,2},$$

The first equation (4.13) corresponds to the usual equation

$$\varepsilon_{11,22} + \varepsilon_{22,11} = 2\varepsilon_{12,12}$$

and remains unaltered in comparison to classical elasticity. The second equation follows immediately from (4.8) since

$$m_{13} = 4\,G\ell^2 \varkappa_{31} = 4\,G\ell^2 \omega_{3,1}\,,$$

$$m_{23} = 4\,G\ell^2 \varkappa_{32} = 4\,G\ell^2 \omega_{3,2}\,.$$

The two last equations (4.13) result from the observation that (cf. Eq. (4.4))

$$\varkappa_{31} = \omega_{3,1} = \frac{1}{2}\left(u_{2,11} - u_{1,12}\right),$$

$$\varepsilon_{12,1} = \frac{1}{2}\left(u_{2,11} - u_{1,12}\right),$$

$$\varepsilon_{11,2} = u_{1,12}$$

and hence

$$\varkappa_{31} - \varepsilon_{12,1} + \varepsilon_{11,2} = 0$$

and similarly

$$\varkappa_{32} + \varepsilon_{12,1} + \varepsilon_{22,1} = 0\,.$$

Let us finally pass to the problem of boundary conditions. If they are posed in displacements the conditions remain extremely simple and require u_1, u_2 and ω_3 to be prescribed along the boundary S of the region occupied by the body. u_3 is identically zero throughout the body, the boundary included, and $u_1 = u_1(x_1, x_2)$, $u_2 = u_2(x_1, x_2)$, and hence the remaining conditions : u_3, ω_1, ω_2 prescribed on the lateral surface of the cylinder are automatically satisfied.

The corresponding number of three boundary conditions expressed in stresses is obtained in the following manner. The six boundary conditions (2. 29), (2. 30)

$$\sigma_{ji} n_j = p_i ,$$

$$\mu_{ji} n_j = q_i ,$$

applied to the lateral surface of the cylinder (Fig. 6) reduce, owing to relations

$$n_3 = 0 \; ; \; \sigma_{3i} = \sigma_{i3} \; ; \; \mu_{12} = \mu_{21} = 0 \; ; \; \mu_{11} = \mu$$

$$\mu_{22} = \mu \; ; \; \mu_{13} = m_{13} \; ; \; \mu_{23} = m_{23}$$

to the following equations

$$p_1 = \sigma_{11} n_1 + \sigma_{21} n_2 ,$$

$$p_2 = \sigma_{12} n_1 + \sigma_{22} n_2 ,$$

$$p_3 = 0 , \tag{4. 14}$$

$$q_1 = \mu n_1 , \; q_2 = \mu n_2 ,$$

$$q_3 = m_{13} n_1 + m_{23} n_2 .$$

If at a given point P of the boundary S a local coordinate system (n, t, g) is introduced with axes normal to the surface, tangent to S and parallel to the x_3-axis, respectively, then (4. 14) transforms to

$$p_{(n)} = \sigma_{(nn)}, \quad p_{(t)} = \sigma_{(nt)}, \quad p_{(g)} = 0,$$

(4.15)

$$q_{(n)} = \mu, \quad q_{(t)} = 0, \quad q_{(g)} = m_{(ng)}.$$

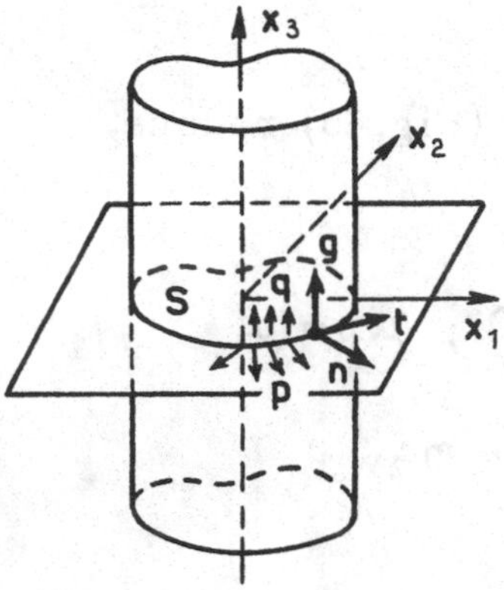

Fig. 6

According to (4.15) the force vector acting on the surface lies in the plane normal to x_3 - in agreement with the classical assumptions; the couple vector tangent to the surface and parallel to the generator is equilibrated by the corresponding component of the couple-stress deviator, whereas the couple vector normal to the surface is constant and equal to the indeterminate couple-stress invariant μ.

The latter condition, as well as the skew-symmetric components $\overset{a}{\sigma}$ of stress can be eliminated from our considerations by the usual procedure of reduced forces, to lead to three boundary conditions in the following form

$$\bar{p}_1 = p_1 = s_{11}n_1 + s_{12}n_2 + \frac{1}{2}(m_{13,1} + m_{23,2})n_2,$$

(4.16a)
$$\bar{p}_2 = p_2 = s_{12}n_1 + s_{22}n_2 - \frac{1}{2}(m_{13,1} + m_{23,2})n_1,$$

$$\bar{q}_3 = q_3 = m_{13}n_1 + m_{23}n_2,$$

$$(\bar{p}_3 = \bar{q}_1 = \bar{q}_2 = 0) . \tag{4.16a}$$

Considering an arbitrary cross-section (x_3 = const.) of the cylinder we are faced with a plane which, owing to our assumptions, does not deform in the process of straining. The forces and couples are transmitted across this plane in the form of stress and couple vectors

$$\begin{aligned} &p_1 = p_2 = 0 , \quad p_3 = \sigma_{33} = s_{33} , \\ &q_1 = \mu_{31} = m_{31} = \eta m_{13} , \\ &q_2 = \mu_{32} = m_{32} = \eta m_{23} , \\ &q_3 = \mu_{33} = \mu , \end{aligned} \tag{4.17}$$

or, using the notations for reduced forces

$$\begin{aligned} &\bar{p}_1 = \bar{p}_2 = 0 , \quad \bar{p}_3 = s_{33} = p_3 , \\ &\bar{q}_1 = q_1 = m_{31} , \quad \bar{q}_2 = q_2 = m_{32} . \end{aligned} \tag{4.18}$$

Summing up the entire plane strain problem can be stated in the following manners.

(a) If the full stress and couple-stress components are taken into consideration, then the problem is not unique owing to the indeterminate value of μ . The number of unknowns (which do not identically vanish) is 18 : 2 displacements u_1, u_2, 1 rotation ω_3, 3 strain and 2 torsion-flexure components :

ε_{11}, ε_{22}, ε_{12} ; $\varkappa_{31}$, $\varkappa_{32}$; 5 force-stresses σ_{11}, σ_{22}, σ_{33}, σ_{12}, σ_{21}, 5 couple stresses μ_{13}, μ_{23}, μ, μ_{31}, μ_{32}. The number of equation is less by one and equals 17; namely : 6 geometric equations (4. 2), (4. 3), (4. 4); 8 "stress-strain" relations (4. 6), (4. 8), (4. 9); 3 equations of motion (or equilibrium) (4. 12). In addition, the boundary conditions (4. 15) involving the indefinite value of μ.

(b) If the skew-symmetric components of σ_{ij} and invariant μ are eliminated from our considerations, the number of unknowns is reduced by two (μ and $\overset{a}{\sigma}_{12}$), whereas the number of governing equations decreases by one (two equations of motion (4. 12a) instead of the three (4. 12)), hence the problem becomes "statically determinate". In addition to that, reduced boundary conditions (4. 16) complete the set of equations.

The problem of integration of the described system of equations can be considerably simplified with the aid of the stress functions introduced by R. D. Mindlin [6] in 1963; this method is most effective in the case when the problem is formulated in stresses and the determination of displacements can be avoided.

Mindlin's assumption generalizing the classical Airy stress function consists in the following representation of the stresses

$$\sigma_{11} = \Phi_{,22} - \Psi_{,12} , \quad \sigma_{22} = \Phi_{,11} + \Psi_{,12} ,$$

$$\sigma_{12} = - \Phi_{,12} - \Psi_{,22} , \quad \sigma_{21} = - \Phi_{,12} + \Psi_{,11} , \tag{4.19}$$

$$\mu_{13} = \Psi_{,1} , \quad \mu_{23} = \Psi_{,2}$$

and, automatically,

$$\sigma_{33} = \nu (\sigma_{11} + \sigma_{22}) = \nu \nabla^2 \Phi$$

$$m_{31} = \eta \Psi_{,1} , \quad m_{32} = \eta \Psi_{,2} . \tag{4.19a}$$

Representation (4.19) has to satisfy the equations of equilibrium and - in absence of all geometric relations - the previously derived equations of continuity (4.13).

Eqs. (4.19) substituted into (4.12) under the assumption that $X_i = 0$, $\ddot{u}_i = 0$ prove to be identically satisfied under the obvious condition of sufficient differentiability of Φ and Ψ. The governing equations for Φ and Ψ are obtained from the compatibility relations. The second relation (4.13) is identically fulfilled; the remaining ones lead to

$$\Phi_{,2222} + \Phi_{,1111} - \nu \nabla^2 \nabla^2 \Phi = - 2 \Phi_{,1122}$$

$$\Psi_{,1} = - 2 \ell^2 (\Phi_{,22} - \Psi_{,12} - \nu \nabla^2 \Phi)_{,2} + \ell^2 (-2 \Phi_{,12} - \Psi_{,22} + \Psi_{,11})_{,1} ,$$

$$\Psi_{,2} = 2 \ell^2 (\Phi_{,11} + \Psi_{,12} - \nu \nabla^2 \Phi)_{,1} - \ell^2 (-2 \Phi_{,12} - \Psi_{,22} + \Psi_{,11})_{,2}$$

or, after simple reductions and rearrangements, to

$$\nabla^2\nabla^2\Phi = 0 ,$$

$$(4.20) \qquad (\Psi - \ell^2\nabla^2\Psi)_{,1} = -2(1-\nu)\ell^2\nabla^2\Phi_{,2} ,$$

$$(\Psi - \ell^2\nabla^2\Psi)_{,2} = 2(1-\nu)\ell^2\nabla^2\Phi_{,1} .$$

The first of equations (4.20) is an obvious result of the remaining ones. On the other hand, differentiation of the second equation of (4.20) with respect to x_1 and the third equation with respect to x_2 leads to a separate equation for Ψ ; this equation is written here together with the first Eq. (4.20):

$$\nabla^2\Psi - \ell^2\nabla^2\nabla^2\Psi = 0 ,$$

$$(4.21) \qquad \nabla^2\nabla^2\Phi = 0 .$$

It is to be remembered that although Eqs. (4.21), being simpler than (4.20), are usually applied to derive the solution of the plane strain problem, this does not absolve us from the necessity of checking whether the original equations (4.20) are satisfied.

In this case, when body forces X_1, X_2 are not absent (body couples being, however, zero), a slightly modified Mindlin's assumption is also applicable, leading so us to non-homogeneous equations. Denoting by U the potential of body forces

$$X_1 = -\frac{\partial U}{\partial x_1} , \qquad X_2 = -\frac{\partial U}{\partial x_2}$$

and assuming the representation

$$\sigma_{11} = U + \Phi_{,22} - \Psi_{,12} ,$$

$$\sigma_{22} = U + \Phi_{,11} + \Psi_{,12} ,$$

(the remaining stresses (4.19) being unchanged), the equations of compatibility take the form

$$\nabla^2 \nabla^2 \Phi = - \frac{1-2\nu}{1-\nu} \nabla^2 U ,$$

$$(\Psi - \ell^2 \nabla^2 \Psi)_{,1} = -2(1-\nu)\ell^2 \nabla^2 \Phi_{,2} - 2(1-2\nu)\ell^2 U_{,2}, \tag{4.22}$$

$$(\Psi - \ell^2 \nabla^2 \Psi)_{,2} = 2(1-\nu)\ell^2 \nabla^2 \Phi_{,1} + 2(1-2\nu)\ell^2 U_{,1} .$$

Here again the first equation results from the two remaining ones. Combining the latter equations with the first one we are led to the non-homogeneous equations

$$\nabla^2 \nabla^2 \Phi = - \frac{1-2\nu}{1-\nu} \nabla^2 U ,$$

$$\nabla^2 \Psi - \ell^2 \nabla^2 \nabla^2 \Psi = 0. \tag{4.23}$$

It is self-evident that the same governing equations could be obtained if the whole problem were formulated in symmetric stress components. Representation (4.19) should then be replaced by

$$s_{12} = - \Phi_{,12} + \frac{1}{2} (\Psi_{,11} - \Psi_{,22})$$

(the remaining stresses $s_{11} = \sigma_{11}$ etc.) and the equilibrium equation should be replaced by its symmetrized counterpart(4.12a).

5. Half-Plane Loaded on the Boundary

Let us consider the typical problem of two-dimensional state of strain, namely a half-plane (or half-space) $x_1 \geq 0$ (Fig. 7) loaded on the finite boundary $x_1 = 0$ by normal and tangential forces.

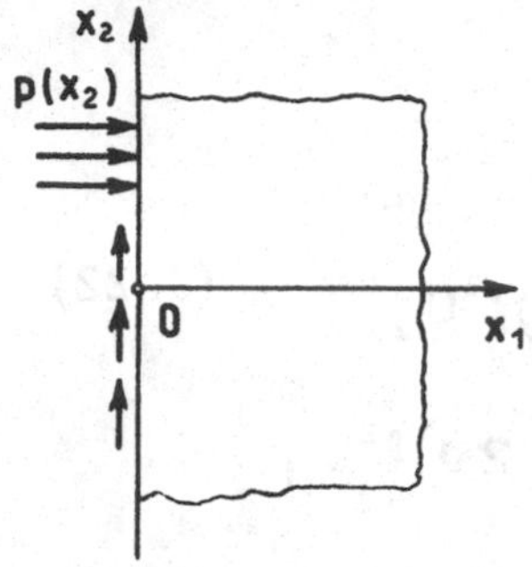

Fig. 7

It is assumed at the beginning of our considerations that the load distribution satisfies all requirements following from the application of the integral Fourier transforms along the x_2-axis (sectional smoothness and absolute integrability in $-\infty < x_2 < +\infty$).

The problem has originally been considered and solved in [2] by R. Muki and E. Sternberg. The solutions should, in addition, satisfy the requirement of vanishing stresses at infinity, whereas the displacements, as it is known from classical elasticity, are allowed to be unbounded at $x_1, x_2 \to \infty$.

To simplify the further considerations let us first assume that the edge $x_1 = 0$ of the plane is loaded by normal forces only $p = p(x_2)$, that the load is symmetric with respect to the origin 0, $p(x_2) = p(-x_2)$, and denote $x_1 = x$, $x_2 = y$; possible generalization of the results to other cases of loadings is simple and does not introduce any essential difficulties.

Owing to formulae (4.19) and (4.20) it is evident that in the case when $\sigma_{11}(0,y)$ is an even function of y, the same applies to functions Φ, σ_{22} and μ_{23}, whereas functions Ψ, σ_{12}, σ_{21} and μ_{13} must be odd functions of y. The Fourier cosine or sine transform can thus be applied to the respective functions. It is known from the general theory of these transforms (cf. e.g. Sneddon [7]) that, on denoting the transforms of $f(y) = f(-y)$ and $g(y) = -g(-y)$ by

$$\bar{f}(\alpha) = \int_0^\infty f(y) \cos \alpha y \, dy ,$$

$$\bar{g}(\alpha) = \int_0^\infty g(y) \sin \alpha y \, dy , \tag{5.1}$$

the inverse transforms are given by

$$f(y) = \frac{2}{\pi} \int_0^\infty \bar{f}(\alpha) \cos \alpha y \, d\alpha ,$$

$$g(y) = \frac{2}{\pi} \int_0^\infty \bar{g}(\alpha) \sin \alpha y \, d\alpha . \tag{5.2}$$

The integrations in (5.1) are performed along the real α - axis.

It is also known that the transform of a derivative of a function $f(y)$ with respect to y equals

(5.3)
$$\int_0^\infty \frac{\partial^{2n} f(y)}{\partial y^{2n}} \cos \alpha y \, dy = (-1)^n \alpha^{2n} \bar{f}(\alpha),$$
$$\int_0^\infty \frac{\partial^{2n+1} g(y)}{\partial y^{2n+1}} \cos \alpha y \, dy = (-1)^n \alpha^{2n+1} \bar{g}(\alpha),$$

(with similar formulae for the sine transforms) provided all derivatives of f or g up to the order by one less than the order of the derivative tend to zero for $y \to \infty$. Since, in our case, we are going to deal with the fourth order differential equations (4.21) for the stress functions Φ and Ψ; it has to be assumed that the third order derivatives of Φ and Ψ vanish at infinity, $|y| \to \infty$. The transform formulae (5.3) do not affect the derivatives with respect to other variables upon which functions f and g might depend, hence, applying these formulae to equations (4.21) we are led to the following fourth order ordinary differential equations

(5.4)
$$\left(\frac{\partial^2}{\partial x^2} - \alpha^2\right) \bar{\Phi} = 0,$$
$$\left(\frac{\partial^2}{\partial x^2} - \alpha^2\right) \bar{\Psi} -$$
$$- \ell^2 \left(\frac{\partial^4}{\partial x^4} - 2\alpha^2 \frac{\partial^2}{\partial x^2} + \alpha^4\right) \bar{\Psi} = 0 .$$

The transformed stresses are expressed with the aid of Eqs. (4.19),

$$\sigma_{11} = -\alpha^2\bar{\Phi} + \alpha\bar{\Psi}_{,x} \;, \quad \sigma_{22} = \bar{\Phi}_{,xx} - \alpha\bar{\Psi}_{,x} \;,$$

$$\sigma_{12} = \alpha\bar{\Phi}_{,x} + \alpha^2\bar{\Psi} \;, \quad \sigma_{21} = \alpha\bar{\Phi}_{,x} - \alpha^2\bar{\Psi} \;, \tag{5.5}$$

$$m_{13} = \bar{\Psi}_{,x} \;, \quad m_{23} = -\alpha\bar{\Psi} \;.$$

The boundary conditions require, according to (4.15), the couple stress m_{13} and force-stress σ_{12} to vanish at $x=0$; the stress σ_{11} assumes the prescribed value of $-\bar{p}(\alpha)$;

$$\frac{\partial}{\partial x}\bar{\Psi}(x,\alpha)\Big|_{x=0} = 0 \;;$$

$$\alpha\,\frac{\partial\Phi(x,\alpha)}{\partial x}\Big|_{x=0} + \alpha^2\bar{\Psi}(0,\alpha) = 0 \;; \tag{5.6}$$

$$-\alpha^2\Phi(0,\alpha) + \alpha\,\frac{\partial\bar{\Psi}}{\partial x}(x,\alpha)\Big|_{x=0} = -\bar{p}(\alpha) \,.$$

The ordinary fourth-order differential equations (5.4) are easily solved to give

$$\bar{\Phi}(x,\alpha) = \left[A(\alpha) + B(\alpha)\,\alpha x\right] e^{-\alpha x} \;,$$

$$\bar{\Psi}(x,\alpha) = C e^{-\alpha x} + D e^{-\sqrt{1+\alpha^2 \ell^2}\, x/\ell} \;.$$

The four integration constants A, B, C, D are found from conditions (5.6) and the original compatibility requirements (4.20). We finally obtain

$$\bar{\Phi}(x,\alpha) = \frac{\bar{p}(\alpha)}{\alpha^2}\left[1 + \alpha x\,\frac{\sqrt{1+\alpha^2\ell^2}}{\Delta(\alpha\ell)}\right]e^{-\alpha x},$$

$$\bar{\Psi}(x,\alpha) = \frac{4(1-\nu)\ell^2\alpha\bar{p}(\alpha)}{\Delta(\alpha\ell)}\left[\frac{\sqrt{1+\alpha^2\ell^2}}{\alpha}\,e^{-\alpha x} - \ell e^{-\sqrt{1+\alpha^2\ell^2}\,x/\ell}\right],$$

with the abbreviated notation

$$\Delta = \Delta(\alpha\ell) = \sqrt{1+\alpha^2\ell^2} + 4(1-\nu)\alpha^2\ell^2\left[\sqrt{1+\alpha^2\ell^2} - \alpha\ell\right].$$

Functions $\bar{\Phi}$ and $\bar{\Psi}$ determined, the stress can be calculated from Eqs. (5.5). Using the additional notation $h(\alpha\ell) = h = \sqrt{1+\alpha^2\ell^2}$ we can write

$$\sigma_{11}(x,y) = -\frac{2}{\pi}\int_0^\infty\Big\{(\Delta + h\alpha x)e^{-\alpha x} + 4(1-\nu)\ell^2 h\alpha^2\left[e^{-hx/\ell} - e^{-\alpha x}\right]\Big\}\frac{\bar{q}(\alpha)\cos\alpha y}{\Delta}\,d\alpha,$$

(5.7a)
$$\sigma_{22}(x,y) = \frac{2}{\pi}\int_0^\infty\Big\{(\Delta - 2h + h\alpha x)e^{-\alpha x} + 4(1-\nu)\ell^2 h\alpha^2\left[e^{-hx/\ell} - e^{-\alpha x}\right]\Big\}\frac{\bar{q}(\alpha)\cos\alpha y}{\Delta}\,d\alpha,$$

$$\sigma_{12} = -\frac{2}{\pi}\int_0^\infty\Big\{(\Delta - h + h\alpha x)e^{-\alpha x} + 4(1-\nu)\ell^2\alpha^2\left[\ell\alpha e^{-hx/\ell} - he^{-\alpha x}\right]\Big\}\frac{\bar{q}(\alpha)\sin\alpha y}{\Delta}\,d\alpha,$$

$$\sigma_{21} = -\frac{2}{\pi}\int_0^\infty \Big\{(\Delta - h + h\alpha x)e^{-\alpha x} +$$

$$+4(1-\nu)\ell h\alpha\left[he^{-hx/\ell} - \ell\alpha e^{-\alpha x}\right]\Big\}\frac{\bar{q}(\alpha)\sin\alpha y}{\Delta}\,d\alpha, \tag{5.7b}$$

$$m_{13} = \frac{8(1-\nu)\ell^2}{\pi}\int_0^\infty \left(e^{-hx/\ell} - e^{-\alpha x}\right)\frac{h\alpha\bar{q}(\alpha)\sin\alpha y}{\Delta}\,d\alpha,$$

$$m_{23} = -\frac{8(1-\nu)\ell^2}{\pi}\int_0^\infty \left(\ell\alpha e^{-hx/\ell} - he^{-\alpha x}\right)\frac{\alpha\bar{q}(\alpha)\cos\alpha x}{\Delta}\,d\alpha.$$

The effective calculation of the stresses depends on the form of function $\bar{p}(\alpha)$, though, in general, it can be performed only numerically, and certain conclusions concerning the asymptotic behavior of the solutions can be drawn. The difficulties in computing the stresses from Eqs. (5.7) result from the structure of the integrand, leading thus to the expressions of the type

$$\int_0^\infty \frac{\exp\left(-\sqrt{\alpha^2 + 1/\ell^2}\,x\right)\bar{q}(\alpha)\cos\alpha x}{\sqrt{1+\alpha^2} + 4(1-\nu)\alpha^2\left[\sqrt{1+\alpha^2} - \alpha\right]}\,d\alpha.$$

Concentrated Force Let us assume that the normal load $p(y)$ applied at the boundary $x = 0$ has the following properties :

$$p_\varepsilon(y) = p_\varepsilon(-y)$$

$$p_\varepsilon(y) = 0 \quad \text{for} \quad y > \varepsilon$$

$$\int_{-\varepsilon}^{+\varepsilon} p_\varepsilon\,dy = P\,;$$

where ε denotes an arbitrary small parameter and P is the constant value of the resultant horizontal force. We are now able to proceed with ε to the limit, $\varepsilon \to 0$, as long as the distribution p_ε satisfies the regularity conditions imposed by the F-transforms. In the limit $\varepsilon \to 0$ function $p_\varepsilon(y)$ assumes the properties of Dirac's delta distribution and hence

$$\bar{p}(\alpha) = \frac{1}{2}\int_{-\infty}^{+\infty} p_\varepsilon(y)\cos\alpha y\, dy = \frac{P}{2}. \tag{5.8}$$

With this value of $\bar{p}(\alpha)$, the normal stress $\sigma_{11}(x,y)$ is expressed by the integral

$$\sigma_{11}(x,y) = -\frac{P}{\pi}\int_0^\infty \Big\{(\Delta + h\alpha x)e^{-\alpha x} +$$

$$+4(1-\nu)\ell^2 h\alpha^2\left(e^{-hx/\ell} - e^{-\alpha x}\right)\Big\}\frac{\cos\alpha y}{\Delta}\,d\alpha. \tag{5.9}$$

It is seen that owing to the introduced notations

$$h = \sqrt{1+\alpha^2\ell^2}$$

$$\Delta = \sqrt{1+\alpha^2\ell^2} + 4(1-\nu)\alpha^2\ell^2\left[\sqrt{1+\alpha^2\ell^2} - \alpha\ell\right]$$

for small values of $\alpha\ell$ (ℓ tending to zero)

$$h \to 1, \quad \Delta \to 1,$$

and the integrand of (5. 9) reduces to the well known integral representation of the classical elasticity solution

$$\sigma_{11} = -\frac{P}{\pi}\int_0^\infty (1+\alpha x)\, e^{-\alpha x} \cos \alpha y \, d\alpha . \tag{5.10}$$

Integrals of the type of (5. 9), (5. 10) become divergent for $x, y \to 0$ which reflects the singular character of elasticity solutions at the point of application of the concentrated force. It is known (see, e. g., Sneddon [7]) that the singular behavior of integrals of the type of (5. 9) at $x, y \to 0$ is characterized by the behavior of the corresponding integrand at large values of α. The asymptotic expansions of the individual terms in (5. 9) are

$$h \sim \alpha \ell + \frac{1}{2\alpha \ell} + 0(\alpha^{-3}),$$

$$\Delta \sim (3-2\nu)\alpha\ell + 0(\alpha^{-1}),$$

$$e^{-hx/\ell} - e^{-\alpha x} = -e^{-\alpha x}\left[\frac{x}{2\alpha\ell^2} + 0(\alpha^{-2})\right],$$

where the symbols $0(\alpha^{-n})$ denote the terms of order α^{-n} and $\alpha \to \infty$. Substituting these asymptotic expansions in Eq. (5. 9) and preserving the highest order terms (in α) in the integral we are led to the approximate expression

$$\sigma_{11}(x, y)\underset{x^2+y^2 \to 0}{\approx} -\frac{P}{\pi}\int_0^\infty \left(1 - \frac{1-2\nu}{3-2\nu}\,\alpha x\right) e^{-\alpha x} \cos \alpha y \, d\alpha,$$

which is easily evaluated to give

$$\sigma_{11}(x,y)\underset{x^2+y^2\to 0}{=} -\frac{2P}{\pi}\frac{x^2+2(1-\nu)xy^2}{(3-2\nu)(x^2+y^2)^2} . \tag{5.11}$$

The remaining terms of asymptotic expansions omitted in (5.11) do not contribute to the singularity at $x^2+y^2=r^2\to 0$.

Comparing this result with the classical solution of the problem (cf., e.g., K. Girkmann [8]),

$$\sigma_{11} = -\frac{2P}{\pi}\frac{x^3}{(x^2+y^2)^2} , \tag{5.12}$$

which is obtained by simple integration of (5.10), we can observe that the order of singularity of the σ_{11} stress at $r\to 0$ is preserved; both expressions (5.11) and (5.12) at $r\to 0$ increase indefinitely as $1/r$, the coefficient of singularity being, however, different in these cases. In the classical elastic problem the coefficient of singularity (as well as the entire solution (5.12)) is independent of ν in accordance with the well-known theorem stating that the elastic moduli do not enter the solutions for stresses provided (a) : the problem is formulated in stresses; b) : no body forces act in the system; c) : the external forces acting on individual boundaries (in the case when the region is multiply connected) form self-equilibrated systems. In the couple stress theory, on the contrary: both the solutions (5.9) and the stress concentration factor depend on Poisson's ratio . What is perhaps even more significant, the stress concentration factor in (5.11) itself is independent of the additional material constant ℓ and hence, with ℓ tending to 0, this factor does not approach the classical value (5.12). This observation, contrary to the fact that the entire formula for stresses (5.9) leads, under $\ell \to 0$, to the classical solution, may seem strange. A similar phenomenon was observed in Section 3 in connection with the singular solution in three dimensions. According to several authors this result reflects certain imperfections of the entire model of micropolar medium and requires further explanation.

On the other hand, the fact that the stress concentration factor remains independent of ℓ is obvious from the point of view of simple dimensional analysis. In the case of such a solution no geometric dimensions of the body represented by a space or semispace can enter the resulting formulae, the body has no finite dimensions. The dimensions of elastic constants G, ν and ℓ are, respectively : Force/length^2, unity and length; the classical stress concentration factor is denoted by K_0 and is dimensionless. Hence, the modified factor K_1 has also to be dimensionless and could be a function of the only material parameters G, ν, ℓ. Though, taking into account their dimensions mentioned above, it is easily established that no combination of G, ν, ℓ can remain dimensionless except when $K_1(G, \nu, \ell)$ is solely a function of ν, $K_1 = K_1(\nu)$

and thus if $K_1 = K_0$, no limiting procedure with $\ell \to 0$ can lead to $K_1 \to K_0$. In the case of the general solution (5.9) this reasoning fails since a new parameter $r = \sqrt{x^2+y^2}$ of the dimension of length appears and the simple combination $r(\ell)$ (or x/ℓ, y/ℓ) proves to be dimensionless. Actually, solution (5.9) can be easily written as a function of these new variables x/ℓ, y/ℓ. In the case of finite regions new parameters describing their size and shape appear in the solution and further dimensionless parameters can be constructed.

To conclude this Section let us quote - without further derivation - the singular (or leading) parts of the solutions of the concentrated force problem derived in [2]. Column A corresponds to the couple-stress solutions, column B represents the (full) solution of the corresponding classical problem. (see next page).

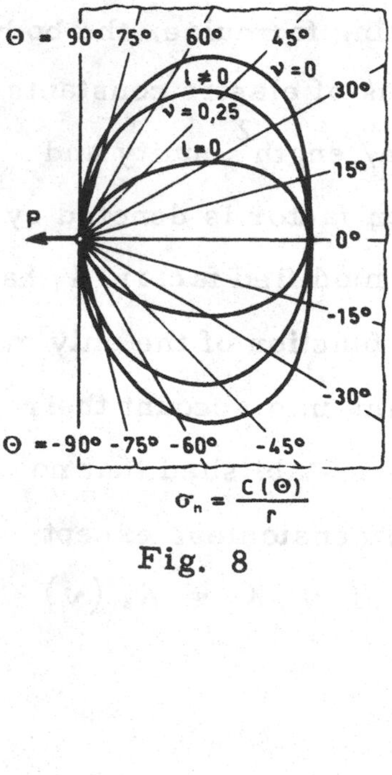

Fig. 8

Symbol f denotes regular parts of the solutions (column A) which remain finite at the origin. It is observed that one of the couple stresses remains finite at the point of application of the force. Fig. 8 shows the variation of the stress concentration factor corresponding to Θ in the couple-stress case and in the classical case. The couple-stress case takes into account two different values of Poisson's ratio : $\nu = 0$ and $\nu = 0{,}25$.

A	B
$\sigma_{11} = -\dfrac{2P}{\pi}\dfrac{x\left[x^2+2(1-\nu)y^2\right]}{(3-2\nu)r^4} + f$	$\sigma_{11} = -\dfrac{2P}{\pi}\dfrac{x^3}{r^4}$
$\sigma_{22} = -\dfrac{2P}{\pi}\dfrac{1-2\nu}{3-2\nu}\dfrac{xy^2}{r^4} + f$	$\sigma_{22} = -\dfrac{2P}{\pi}\dfrac{xy^2}{r^4}$
$\sigma_{12} = \dfrac{2P}{\pi}\dfrac{1-2\nu}{3-2\nu}\dfrac{x^2y}{r^4} + f$	$\sigma_{12} = \sigma_{21} =$
$\sigma_{21} = -\dfrac{2P}{\pi}\dfrac{y\left[x^2+2(1-\nu)y^2\right]}{(3-2\nu)r^4} + f$	$= -\dfrac{2P}{\pi}\dfrac{x^2y}{r^4}$
$m_{13} = f$	$m_{13} = 0$
$m_{23} = -\dfrac{2P}{\pi}\dfrac{1-\nu}{3\;\;2\nu}\log r + f$	$m_{23} = 0$

6. Circular Holes and Inclusions

Let us pass to the consideration of certain plane problems of micropolar elasticity in which the stress concentrations are not caused by a singular load distribution but by a special form of the loaded body. The simplest and perhaps the most important and widely known example of such a stress concentration is encountered when a plane sheet or plate subject to arbitrary plane state of strain (or stress) contains an inclusion made of a different material. Two limiting cases can be considered here : if the material of the inclusion is assumed to be perfectly rigid then the boundary conditions at the interface between the original material and the inclusion are very simple; the inclusion does not deform and behaves as a rigid body; the displacements of the original material at the boundary of the inclusion either are zero or reflect a possible rigid translation and rotation. It is assumed that both materials are glued together fulfilling thus the continuity conditions before and after deformation.

The other limiting case is obtained when the elastic moduli of the inclusion E, G are assumed to be zero - in other words we are dealing with a hole or cavity in an elastic medium. The conditions at the boundary of the hole are again extremely simple; the surface stress vector either vanishes (if the cavity is free of internal pressure) or assumes prescribed values (if the cavity is loaded from inside). In addition to these limiting cases, all intermediate cases of various inclusion material properties are possible. All these cases (except the trivial one when both materials are identical) lead to certain variations of the original stress distribution. In certain cases a definite stress concentration at the boundary can be observed, a phenomenon of a great technical importance which often proves to be responsible for the material fracture, especially when the boundaries of the cavities or inclusions contain sharp corners.

Let us start with the analysis of the stress concentration problems of circular holes and inclusions and evaluate the influence of couple-stresses upon the stress concentration factors.

Circular Hole (Cylindrical Cavity) The couple-stress solution for uniformly loaded plate containing a circular hole (plane stress) or an elastic space containing a spherical cavity (plane strain) was originally considered by R. D. Mindlin [6] in 1963. The problem can - according to Section 4 of this paper - be reduced to the solution of the system (4. 21) of differential equations for the stress functions Φ and Ψ ,

$$\begin{aligned} &\nabla^2 \nabla^2 \Phi = 0, \\ &\nabla^2 \Psi - \ell^2 \nabla^2 \nabla^2 \Psi = 0, \end{aligned} \qquad (6.1)$$

and the additional conditions

$$(6.2)\qquad \begin{aligned} (\Psi - \ell^2 \nabla^2 \Psi)_{,1} &= -2(1-\nu)\ell^2 \nabla^2 \Phi_{,2}\,, \\ (\Psi - \ell^2 \nabla^2 \Psi)_{,2} &= 2(1-\nu)\ell^2 \nabla^2 \Phi_{,1}\,, \end{aligned}$$

the stresses σ_{ij} and μ_{ij} to be determined from (4.19)

$$\sigma_{11} = \Phi_{,22} - \Psi_{,12}\,, \qquad \sigma_{22} = \Phi_{,11} + \Psi_{,12}\,,$$

$$\sigma_{12} = -\Phi_{,12} - \Psi_{,22}\,, \qquad \sigma_{21} = -\Phi_{,12} + \Psi_{,11}\,,$$

$$\mu_{13} = \Psi_{,1}\,, \qquad \mu_{23} = \Psi_{,2}$$

In a cylindrical coordinate system which is more suitable in the case considered here (Fig. 9) the corresponding formulae take the form

$$(6.3)\qquad \begin{aligned} \sigma_{rr} &= \frac{1}{r}\frac{\partial \Phi}{\partial r} + \frac{1}{r^2}\frac{\partial^2 \Phi}{\partial \Theta^2} - \frac{\partial}{\partial r}\left(\frac{1}{r}\frac{\partial \Psi}{\partial \Theta}\right), \\ \sigma_{\Theta\Theta} &= \frac{\partial^2 \Phi}{\partial r^2} + \frac{\partial}{\partial r}\left(\frac{1}{r}\frac{\partial \Psi}{\partial \Theta}\right), \\ \sigma_{\Theta\Theta} &= -\frac{\partial}{\partial r}\left(\frac{1}{r}\frac{\partial \Phi}{\partial \Theta}\right) - \frac{1}{r}\frac{\partial \Psi}{\partial r} - \frac{1}{r^2}\frac{\partial^2 \Psi}{\partial \Theta^2}\,, \\ \sigma_{\Theta r} &= -\frac{\partial}{\partial r}\left(\frac{1}{r}\frac{\partial \Phi}{\partial \Theta}\right) + \frac{\partial^2 \Psi}{\partial r^2}\,, \\ \mu_{rz} &= \frac{\partial \Psi}{\partial r}\,, \\ \mu_{\Theta z} &= \frac{1}{r}\frac{\partial \Psi}{\partial \Theta}\,. \end{aligned}$$

The governing equations (6.1) remain unchanged provided the symbol ∇^2 is written in plane polar coordinates

$$\nabla^2 \equiv \frac{\partial^2}{\partial r^2} + \frac{1}{r}\frac{\partial}{\partial r} + \frac{1}{r^2}\frac{\partial^2}{\partial \Theta^2} ,$$

whereas Eqs. (6.2) assume the form

$$\begin{aligned} \frac{\partial}{\partial r}(\Psi - \ell^2\nabla^2\Psi) &= -2(1-\nu)\ell^2\frac{1}{r}\frac{\partial}{\partial\Theta}\nabla^2\Phi , \\ \frac{1}{r}\frac{\partial}{\partial\Theta}(\Psi - \ell^2\nabla^2\Psi) &= 2(1-\nu)\ell^2\frac{\partial}{\partial r}\nabla^2\Phi . \end{aligned} \tag{6.4}$$

The boundary conditions at $r = a$ (Fig. 9) are reduced to the requirement that all components of the force-stress and couple-stress vanish.

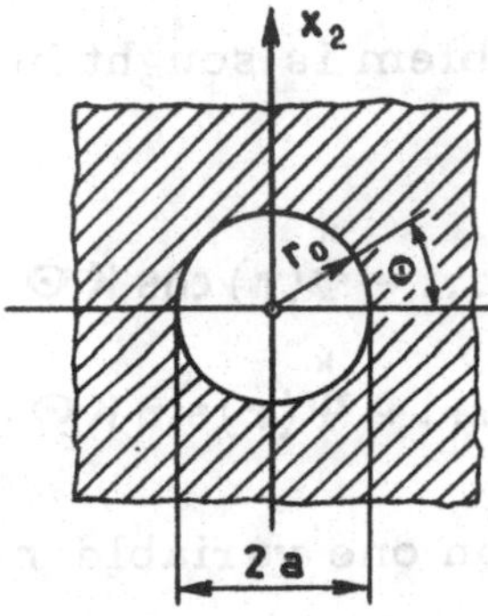

Fig. 9

At a certain distance of the hole the stresses should approach the prescribed value depending upon the way the body is loaded at infinity. For simplicity it will be assumed that the plate extends to infinity and with increasing r the state of stress becomes uniform, i.e. the plate is loaded at infinity by force-stress vectors

$$p_i = \sigma_{ji}\, n_j$$

σ_{ij} being independent of x, y, r, Θ. If, for instance, the plate is subject to uniform and uni-axial tension, then at infinity the stresses approach the values

$$\sigma_{11} = p\,,\quad \sigma_{22} = \sigma_{21} = \sigma_{12} = \mu_{13} = \mu_{23} = 0$$

or, in polar coordinates

$$\sigma_{rr} = p\cos^2\Theta\,,\quad \sigma_{\Theta\Theta} = p\sin^2\Theta\,,$$

$$\sigma_{r\Theta} = \sigma_{\Theta r} = -\frac{p}{2}\sin 2\Theta\,,\quad \mu_{rz} = \mu_{\Theta z} = 0$$

which corresponds to the loading of the boundary by forces

$$p_1 = p n_1\,,\quad p_2 = q_1 = q_2 = 0\,.$$

Following the method applied in classical elasticity, the solution of the posed boundary value problem is sought in the form of a trigonometric series

$$(6.5)\quad \begin{aligned} \Phi(r,\Theta) &= \overset{0}{\Phi}(r) + \overset{1}{\Phi}(r)\cos\Theta + \ldots + \overset{K}{\Phi}(r)\cos K\Theta\,,\\ \Psi(r,\Theta) &= \qquad\qquad \overset{1}{\Psi}(r)\sin\Theta + \ldots + \overset{K}{\Psi}(r)\sin K\Theta\,, \end{aligned}$$

with functions $\overset{i}{\Phi}$, $\overset{i}{\Psi}$ depending solely on one variable r. Expressions (6.5) substituted in (6.1), ordinary differential equations result

$$\left(\frac{d^2}{dr^2} + \frac{1}{r}\frac{d}{dr} - \frac{K^2}{r^2}\right)^2 \overset{K}{\Phi}(r) = 0\,,$$

$$\left[\left(\frac{d^2}{dr^2} + \frac{1}{r}\frac{d}{dr} - \frac{K^2}{r^2}\right) - \ell^2\left(\frac{d^2}{dr^2} + \frac{1}{r}\frac{d}{dr} - \frac{K^2}{r^2}\right)^2\right]\overset{K}{\Psi} = 0\,.$$

The equation for Φ is unchanged when compared to the classical Airy stress function and it is inegrated as usual. Equation for Ψ can be written as

$$\left(\frac{d^2}{dr^2}+\frac{1}{r}\frac{d}{dr}-\frac{k^2}{r^2}\right)\left(\ell^2\frac{d^2}{dr^2}+\ell^2\frac{1}{r}\frac{d}{dr}-1-\frac{\ell^2k^2}{r^2}\right)\overset{k}{\Psi}=0.$$

The solution for $\overset{k}{\Psi}$ is now written as a combination of a solution of a harmonic equation and of the equation

$$\ell^2\frac{d^2\overset{k}{\Psi}}{dr^2}+\ell^2\frac{1}{r}\frac{d\overset{k}{\Psi}}{dr}-\left(1+\frac{\ell^2k^2}{r^2}\right)\overset{k}{\Psi}=0$$

which, with the aid of the new variable $\rho=r/\ell$ can be written as

$$\frac{d^2\overset{k}{\Psi}}{d\rho^2}+\frac{1}{\rho}\frac{d\overset{k}{\Psi}}{d\rho}-\left(1+\frac{k^2}{\rho^2}\right)\overset{k}{\Psi}=0 \qquad (6.6)$$

and is easily recognized as a Bessel differential equation. The integral of (6.6) is written as

$$\overset{k}{\Psi}(r)=\overset{k}{A}\,I_k\left(\frac{r}{\ell}\right)+\overset{k}{B}\,K_k\left(\frac{r}{\ell}\right),$$

with I_k, K_k denoting the modified Bessel functions of the k-th order.

Now, the further procedure is rather simple: functions $\overset{k}{\Phi}$ and $\overset{k}{\Psi}$ are written as combinations of exponential, logarithmic and Bessel functions; the number of constants of integration is reduced by the requirement that the stresses remain finite at infinity (which excludes positive powers or r and functions I_k). It is finally found that Φ and Ψ should be assumed in the form

$$\Phi(r,\Theta) = \frac{1}{4}\, p r^2 (1 - \cos 2\Theta) + A \log r + \left(\frac{B}{r^2} + C\right) \cos 2\Theta,$$

(6.7)

$$\Psi(r,\Theta) = \left[\frac{D}{r^2} + E K_2\left(\frac{r}{\ell}\right)\right] \sin 2\Theta .$$

The remaining constants of integration $A, \ldots, E$ are found from the boundary conditions at the edge of the circular hole

$$\sigma_{rr} = \sigma_{r\Theta} = \mu_{rz} = 0$$

and from the additional requirements (6.4). These conditions lead to the result

$$A = -\frac{pa^2}{2}, \quad B = -\frac{pa^4}{4}\frac{1-F}{1+F}, \quad C = \frac{pa^2}{2}\frac{1}{1+F},$$

(6.8)

$$D = \frac{4(1-\nu)pa^2\ell^2}{1+F}, \quad E = -\frac{pa\ell F}{(1+F)K_1(a/\ell)}$$

$$F = \frac{8(1-\nu)}{4 + \dfrac{a^2}{\ell^2} + \dfrac{2a}{\ell}\dfrac{K_0(a/\ell)}{K_1(a/\ell)}} .$$

The classical elasticity solution of the corresponding problem (known as the Kirsch problem, cf. e.g. Timoshenko-Goodier [9]) leads to the Airy stress function

$$\Phi(r,\Theta) = \frac{1}{4} pr^2(1-\cos 2\Theta) - \frac{pa^2}{2} \log r + \left(-\frac{pa^4}{4}\frac{1}{r^2} + \frac{pa^2}{2}\right)\cos 2\Theta . \qquad (6.9)$$

Comparing (6.7) and (6.9) we can observe that for $\ell \longrightarrow 0$ both solutions coincide, since

$$\lim \frac{K_0(u)}{K_1(u)} = 1 ,$$

$$\lim_{\ell \to 0} F = \lim_{\ell \to 0} E = \lim_{\ell \to 0} D = 0 ;$$

$$A = -\frac{pa^2}{2} , \quad \lim_{\ell \to 0} B = -\frac{pa^4}{4} , \quad \lim C = \frac{pa^2}{2} .$$

Formula (6.9) represents the closed form solution of the "generalized Kirsch problem" which is rather rare in the couple-stress theory. The discussion of the full solution (6.7), (6.8) in the entire region of the plate would be troublesome, though (and hence we reduce the discussion to the analysis of the area of the stress concentration). The classical solution gives the circumferential stress $\sigma_{\Theta\Theta}$ at the hole boundary equal to

$$\sigma_{\Theta\Theta} = p\,(1-\cos 2\Theta),$$

which becomes maximum at

$$\Theta = \pm \frac{\pi}{2},$$

$$\sigma_{\Theta\Theta}^{max} = 3p = Cp,$$

the stress concentration factor being equal to 3. In the couple-stress solution the maximum hoop stress occurs at the same point of the boundary of the hole, the maximum value of $\sigma_{\Theta\Theta}$ being

$$\sigma_{\Theta\Theta}^{max} = 3p \frac{1 + \frac{1}{3}F}{1 + F} = C\left(\frac{a}{\ell}, \nu\right)p. \tag{6.10}$$

The stress concentration factor evidently depends on F, and hence on a/ℓ and ν, which follows from Eq. (6.8).

The variation of the stress concentration factor with the ratio ℓ/a and ν has been evaluated by Mindlin [6] and is shown in Fig. 10.

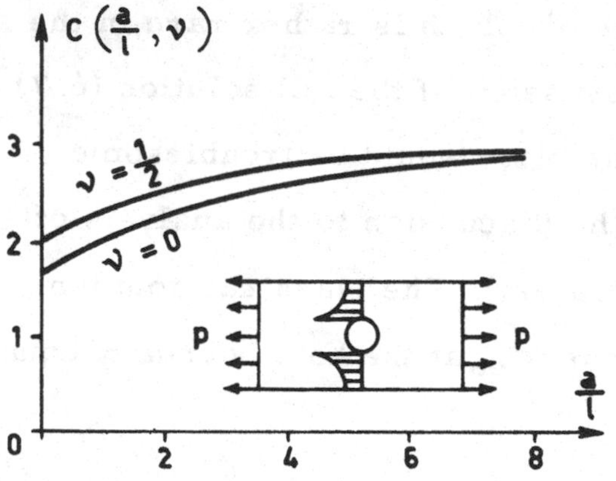

Fig. 10

A very similar procedure can be applied when the plate is loaded on all edges : by horizontal forces p on vertical edges and by vertical forces q on horizontal edges. When $p = -q$, the so-called state of pure shear results (Fig. 11) and the stress concentration occurs at four points $\Theta = 0, \pm\frac{\pi}{2}, \pi$ (two maxima and two minima), $\sigma_{\Theta\Theta}\left(a, \pm\frac{\pi}{2}\right) = -\sigma_{\Theta\Theta}\left(a, \frac{\pi}{2} \pm \frac{\pi}{2}\right)$. The corresponding stress concentration factor

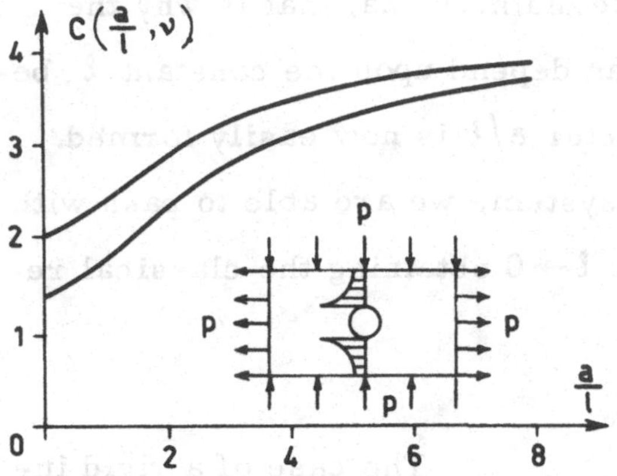

Fig. 11

$$|\sigma_{\Theta\Theta}(a, \Theta)|^{max} = c\left(\frac{a}{\ell}, \nu\right) p$$

equals here

$$c\left(\frac{a}{\ell}, \nu\right) = \frac{4}{1+F} \tag{6.11}$$

and is represented in Fig. 11.

Both particular solutions reveal certain significant and important properties. First of all, a considerable decrease of the stress concentration factor is observed, especially for small values of the ration a/ℓ i.e. when either the additional material conctant ℓ is large or the hole diameter is very small. The distinct effect of the scale appears here : the stress concentration factor depends on the size of the hole, contrary to the classical elasticity solution.

The second observation is the following : the system, in contrast to the previously discussed ones, is characterized by the usual material constants E, ν, ℓ and the additional geometric parameter, the hole diameter 2a; that is why the stress concentration factor can depend upon the constant ℓ, because a dimensionless parameter a/ℓ is now easily formed. Owing to this property of the system, we are able to pass with the final solutions to the limit $\ell \to 0$ obtaining the classical results $C = 3$ and $C = 4$.

Rigid Circular (Cylindrical) Inclusion The case of a rigid inclusion in the plate subject to uni-axial extension is discussed in the paper by Banks and Sokolowski [10]. The procedure is similar to the one presented above, the stress functions have the same form (6.7); the integration constants are calculated from the two conventional conditions

$$u_r(a,\Theta) = u_\Theta(a,\Theta) = 0 \tag{6.12}$$

and the third one : rotation ω_z at the boundary of the inclusion has to vanish owing to perfect joining of both materials. Since from (6.12) it is evident that $\partial u_r / \partial \Theta = 0$ at $r = 0$, this requirement reduces to

$$\left. \frac{\partial u_\Theta(r,\Theta)}{\partial r} \right|_{r=a} = 0 . \tag{6.13}$$

Conditions (6.12), (6.13) lead to the following values of the constants of integration $A \ldots F$ in (6.7)

$$A = \frac{pa^2}{2}(1-2\nu), \quad B = \frac{pa}{2}\left[(3-2\nu) + \frac{a}{2\ell}\frac{K_0(a/\ell)}{K_1(a/\ell)}\right]\frac{1}{F},$$

$$C = -\frac{pa}{2}\left[2 + \frac{a}{\ell}\frac{K_0(a/\ell)}{K_1(a/\ell)}\right]\frac{1}{F},$$

$$D = -4p(1-\nu)a^2\ell^2\left[2 + \frac{a}{\ell}\frac{K_0(a/\ell)}{K_1(a/\ell)}\right]\frac{1}{F},$$

$$E = \frac{8p(1-\nu)a\ell}{K_1(a/\ell)}$$

with

$$F = 2(1-2\nu) + (3-4\nu)\frac{a}{\ell}K_0(a/\ell)/K_1(a\ \ell).$$

The maximum value of normal stresses occurs at the ends of the horizontal diameter $\Theta = 0, \pi$,

$$\sigma_{rr}^{max} = \frac{p}{2}\left[(3-2\nu) + \frac{2(3-2\nu) + \frac{a}{\ell}K_0\left(\frac{a}{\ell}\right)/K_1\left(\frac{a}{\ell}\right)}{2(1-2\nu) + (3-4\nu)\frac{a}{\ell}K_0\left(\frac{a}{\ell}\right)/K_1\left(\frac{a}{\ell}\right)}\right]. \tag{6.14}$$

The classical result obtained by Goodier has the form

$$\sigma_{rr}^{max} = \frac{p}{2}\frac{2(5-4\nu)(1-\nu)}{3-4\nu}. \tag{6.15}$$

With $\ell \to 0$, $\lim K_0\left(\frac{a}{\ell}\right)/K_1\left(\frac{a}{\ell}\right) = 1$, the expression in brackets of (6.14) transforms to

$$\frac{p}{2}\left[3-2\nu+\frac{1}{3-4\nu}\right]=\frac{p}{2}\;\frac{2(5-4\nu)(1-\nu)}{3-4\nu}$$

and thus both formulae coincide. For $\ell \neq 0$ (6.14) leads to higher stress concentration factors which is shown in the diagram (Fig. 12). Here both the classical and the couple-stress results depend on Poisson's ratio ν (the boundary condition at the inclusion is of a geometric character). Though, in contrast to our previous results, the couple-stresses lead to an increase (not decrease) of the stress concentration; a similar difference will be observed by comparing the influence of couple stresses on stress concentrations connected with cracks and rigid blocks in micropolar bodies.

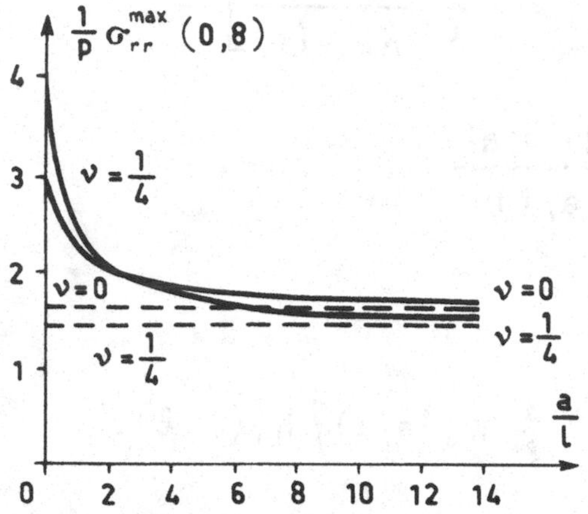

Fig. 12

Application of Complex Variable Functions The well-known and widely used in classical elasticity methods of complex variable functions can also be extended to the analysis of the couple-stress theory. This method is based on the simple observation that the real and imaginary parts of analytic functions

$f(z) = u(z)+iv(z)$ of a complex variable $z = x + iy$ are harmonic functions of x and y

$$\frac{\partial^2 f(x+iy)}{\partial x^2} = \frac{\partial^2 u(x+iy)}{\partial x^2} + i\,\frac{\partial^2 v(x+iy)}{\partial x^2} =$$

$$= -\frac{\partial^2 u}{\partial y^2} - i\,\frac{\partial^2 v}{\partial y^2}\,,$$

$$\nabla^2 u(x,y) = \nabla^2 v(x,y) = 0$$

and on the Goursat representation of a biharmonic function $F(x,y)$,

$$F = \Re e\left[z\varphi(z) + \chi(z)\right] =$$

$$= \frac{1}{2}\left[\bar{z}\,\varphi(z) + z\,\overline{\varphi(z)} + \chi(z) + \overline{\chi(z)}\right] \qquad (6.16)$$

where $\varphi(z)$, $\chi(z)$ are analytic functions of $z = x + iy$. $\bar{z}$ denotes the complex conjugate number $x - iy$.

Writing the Airy stress function in the form (6.16) (in absence of body forces), the stresses and displacements in classical two-dimensional elasticity can be written in terms of the two analytic functions φ and χ ; the most complete description and derivation of the method can be found in the book by I. N. Muschelishvili [12],

$$\sigma_{xx}+\sigma_{yy} = 2\left[\varphi'(z) + \overline{\varphi'(z)}\right],$$

$$\sigma_{yy} - \sigma_{xx} + 2i\sigma_{xy} = 2\left[\bar{z}\,\varphi''(z) + \chi''(z)\right],$$

$$u_x + i u_y = \frac{1}{2G}\left[\varkappa\,\varphi(z) - z\,\overline{\varphi'(z)} - \overline{\chi'(z)}\right].$$

In polar coordinates these formulae read

$$\sigma_{rr} + \sigma_{\Theta\Theta} = 2\left[\varphi'(z) + \overline{\varphi'(z)}\right],$$

$$\sigma_{\Theta\Theta} - \sigma_{rr} + 2i\sigma_{r\Theta} = 2e^{2i\Theta}\left[\bar{z}\,\varphi''(z) + \chi''(z)\right],$$

$$2G\,(u_r + iu_\Theta) =$$

$$= \left[\varkappa\,\varphi(z) - z\,\overline{\varphi'(z)} - \overline{\chi'(z)}\right]e^{-i\Theta}.$$

This method can be, after considerable modifications, applied to the plane problems of the couple-stress theory which as it is known from Section 4 - is not based on one biharmonic Airy stress function but on two functions Φ and Ψ, Eqs. (4.21). Three functions of the complex variable have to be introduced here; functions $\varphi(z)$ and $\chi(z)$ are the suitably chosen analytic function of $z = x + iy$, and function $\Omega(z)$ represents a solution of equation

$$\Omega - \ell^2\nabla^2\Omega = 0 \tag{6.17}$$

The components of the stresses and displacements are expressed in terms of φ, χ and Ω as follows :

(6.18a)

$$\sigma_{rr} + \sigma_{\Theta\Theta} = 2\left[\varphi'(z) + \overline{\varphi'(z)}\right],$$

$$\sigma_{\Theta r} - \sigma_{r\Theta} = \frac{1}{\ell^2}\Omega,$$

$$\sigma_{\theta\theta} - \sigma_{rr} + i(\sigma_{r\theta} + \sigma_{\theta r}) =$$

$$= 2e^{2i\theta}\left[\bar{z}\varphi''(z) + 8(1-\nu)\ell^2\varphi'''(z) + \chi''(z) + 2i\Omega''(z)\right],$$

$$\mu_{rz} - i\mu_{\theta z} = 2e^{i\theta}\left[-4i(1-\nu)\ell^2\varphi''(z) + \Omega'(z)\right], \tag{6.18b}$$

$$2G(u_r + iu_\theta) = (3-4\nu)\varphi(z) - z\overline{\varphi'(z)} -$$

$$-8(1-\nu)\ell^2\overline{\varphi''(z)} - \chi'(z) + 2i\frac{\partial\Omega}{\partial\bar{z}}$$

$$4G\omega_z = \frac{\Omega}{\ell^2} - 4i(1-\nu)\left[\varphi'(z) - \overline{\varphi'(z)}\right].$$

The boundary conditions yield the additional data concerning the functions sought for. Two fundamental sets of boundary conditions can be assumed here.

(1). Dynamic boundary conditions. The vectors of force and couple-stresses are prescribed on the boundary. Denoting by N , T and M the normal and tangential components of the force-stress vector and the axial components of the couple-stress vector, respectively, applied to a circular boundary, this set of boundary conditions can be written in the form

$$\varphi'(z) + \overline{\varphi'(z)} - e^{2i\theta}\left[\bar{z}\varphi''(z) + 8(1-\nu)\ell^2\varphi'''(z) + \right.$$

$$\left. + \chi''(z) + 2i\frac{\partial^2\Omega}{\partial z^2}\right] + \frac{i}{2\ell^2}\Omega = N - iT \tag{6.19}$$

$$2\,\mathrm{Re}\left\{e^{i\theta}\left[-4i(1-\nu)\ell^2\varphi''(z) + \Omega'(z)\right]\right\} = M.$$

(2) Geometric boundary conditions. The vector of displacement and rotation is prescribed on the boundary; denoting its components by U_r, U_θ and Ω_z, these conditions are written in the form

$$(3-4\nu)\varphi(z) - z\overline{\varphi'(z)} - 8(1-\nu)\ell^2\overline{\varphi''(z)} -$$

$$-\chi'(z) + 2i\Omega' = 2G(U_r + iU_\theta), \tag{6.20}$$

$$\frac{\Omega}{\ell^2} - 8(1-\nu)\,\mathfrak{Im}\left[\varphi'(z)\right] = 4G\Omega_z .$$

Using representation (6.18) the two analytic functions φ, χ, and function Ω satisfying (6.17),a number of micropolar elasticity problem can be solved. Numerous solutions of this type concerning the question of finite stress concentrations around holes and inclusions were solved and presented in the book by G.N. Savin [13] in the chapter "influence of asymmetry of the stress tensor upon the stress distribution at holes"; see also papers [14, 15].

The problems of circular holes and inclusions, when solved by the complex potential method, lead obviously to solutions identical with those presented above. E.g., the problem of a circular hole in a plate subject to simple extension is solved if functions φ, χ and Ω are assumed in the form

$$\varphi(z) = \frac{p}{4}z + \frac{c}{z},$$

$$\chi(z) = -\frac{p}{4}z^2 + A\log z + \frac{B}{z^2},$$

$$\Omega(r,\Theta) = D K_2\left(\frac{r}{\ell}\right)\sin 2\Theta.$$

The stresses following from these assumptions coincide with those previously calculated. In the book [13] a diagram is plotted which shows the variation of σ_{rr} and $\sigma_{\Theta\Theta}$ with increasing r in the vicinity of the hole (Fig. 13). Solid lines correspond to the couple-stress solution, $\nu = 0{,}25$, $a/\ell = 3$, dotted lines correspond to the classical, couple-stress free solution. It is seen here that in addition to a certain reduction of the $\sigma_{\Theta\Theta}$ - stress concentration factor, the existence of couple-stresses slightly influences the stress distribution at larger distances from the hole. The influence is however limited to distances not exceeding several lengths ℓ.

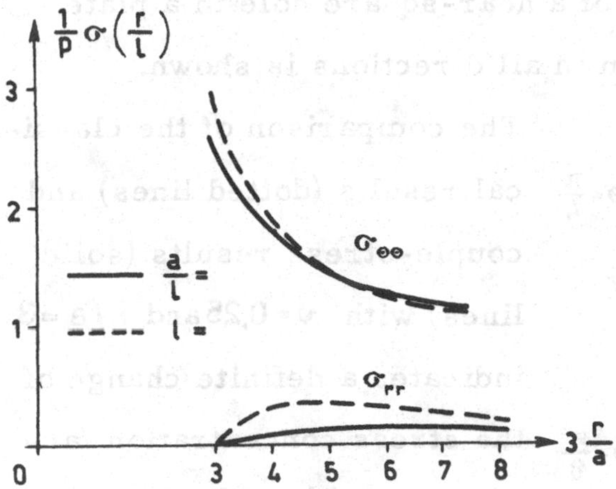

Fig. 13

G. N. Savin and A. N. Guz discussed and solved in paper [16] a more general problem of stress concentration around holes of non-circular forms; the boundary of the hole is described by parametric equations

$$x = R\left(\cos\Theta + \varepsilon\sum_1^n c_k \cos k\vartheta\right),$$

$$y = R\left(\sin\Theta - \varepsilon \sum_{1}^{n} c_k \sin k\vartheta\right),$$

where $\varepsilon \ll 1$, a small, constant parameter characterizes the deviation of the contour from the circle. Elliptic, near-triangular and near-square (with rounded edges) holes were considered and approximate results were plotted. In Fig. 14 the stress $\sigma_{\Theta\Theta}$ at the boundary of a near-square hole in a plate subject to uniform extension in all directions is shown.

The comparison of the classical results (dotted lines) and couple-stress results (solid lines) with $\nu = 0{,}25$ and $R/a = 3$ indicates a definite change of the stress concentration factor; a considerable decrease can be observed at the sharp edge of the hole.

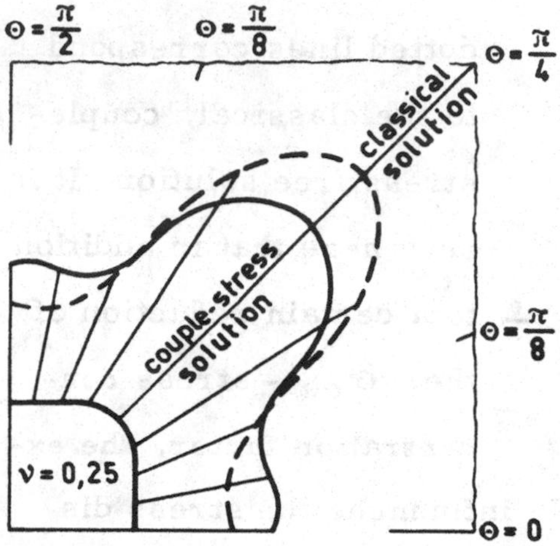

Fig. 14

7. Torsion of Prismatical Bars

A very typical problem of classical elasticity consists in the consideration of a prismatical (cylindrical) bar free of stresses on its lateral surface and loaded on its ends by suitably distributed forces which can be reduced to a torque twisting the cylinder about the axis. The Saint Venant theory of torsion is based on the assumption that the cylindrical "slices" are subject to simple rotation and do not deform in the plane perpendicular to the axis and that the axial displacement of the bar is independent of the coordinate $x_3 = z$, $u_3 = u_3(x_1, x_2)$ and represents the so-called "warping" of the cross-section.

It was shown in [17] that similar assumptions can be made in the couple-stress theory of torsion. These assumptions concerning the state of displacement lead however to a more complicated equation governing the warping function as well as to more complicated boundary conditions.

The problem of torsion of a circular cylinder was discussed earlier by Koiter [1].

Let us consider a cylinder of arbitrary cross-section the contour of which will be assumed smooth (without sharp corners). To solve the problem let us assume that the displacements in a twisted bar can be expressed in the form (Fig. 15)

$$u = -\tau z y\ , \quad v = \tau z x\ , \tag{7.1}$$
$$w = \tau\varphi(x,y),$$

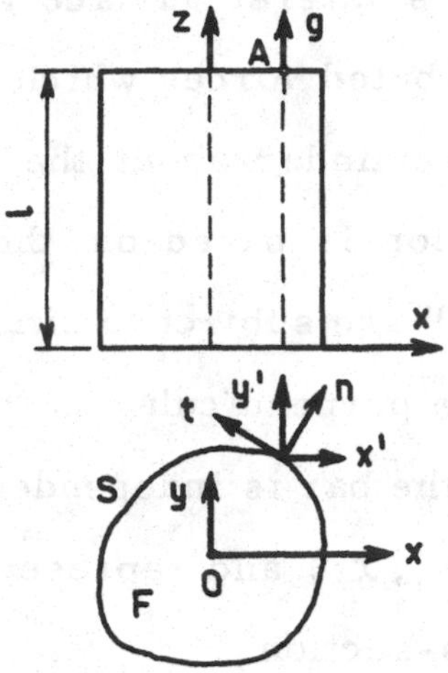

Fig. 15

with constant τ denoting the specific angle of twist and function $\varphi(x,y)$ describing the warping of the cross-section. Inserting (7.1) into the expressions for ε_{ij} and $\varkappa_{ij}$ we are led to

$$\varepsilon_{xx} = \varepsilon_{yy} = \varepsilon_{zz} = \varepsilon_{xy} = \varepsilon_{yx} = 0$$

(7.2)

$$\varepsilon_{xz} = \varepsilon_{zx} = \frac{\tau}{2}\left(\frac{\partial\varphi}{\partial x} - y\right), \quad \varepsilon_{yz} = \varepsilon_{zy} = \frac{\tau}{2}\left(\frac{\partial\varphi}{\partial y} + x\right),$$

$$\varkappa_{xx} = \frac{\tau}{2}\left(\frac{\partial^2\varphi}{\partial x\,\partial y} - 1\right), \quad \varkappa_{yy} = -\frac{\tau}{2}\left(\frac{\partial^2\varphi}{\partial x\,\partial y} + 1\right), \quad \varkappa_{zz} = \tau,$$

$$\varkappa_{xy} = -\frac{\tau}{2}\frac{\partial^2\varphi}{\partial x^2}, \quad \varkappa_{yx} = \frac{\tau}{2}\frac{\partial^2\varphi}{\partial y^2}. \tag{7.3}$$

With the aid of the constitutive equations the matrices of tensors σ_{ij} and μ_{ij} can be written

$$\begin{aligned} &\sigma_{xx} = 0, \quad \sigma_{xy} = \overset{a}{\sigma}_{xy}, \quad \sigma_{xz} = G\tau(\varphi_{,x} - y) + \overset{a}{\sigma}_{xz}, \\ &\sigma_{yz} = -\overset{a}{\sigma}_{xy}, \quad \sigma_{yy} = 0, \quad \sigma_{yz} = G\tau(\varphi_{,y} + x) + \overset{a}{\sigma}_{yz}, \\ &\sigma_{zx} = G\tau(\varphi_{,x} - y) - \overset{a}{\sigma}_{xz}, \quad \sigma_{zy} = G\tau(\varphi_{,y} + x) - \overset{a}{\sigma}_{yz}, \quad \sigma_{zz} = 0 \end{aligned} \tag{7.4}$$

$$\begin{aligned} &\mu_{xx} = 2B\tau(\varphi_{,xy} - 1) + \mu, \quad \mu_{xy} = -2B\tau\varphi_{,xx}, \quad \mu_{xz} = 0 \\ &\mu_{yz} = 2B\tau\varphi_{,yy}, \quad \mu_{yy} = -2B\tau(\varphi_{,xy} + 1) + \mu, \quad \mu_{yz} = 0 \\ &\mu_{zx} = 0, \quad \mu_{zy} = 0, \quad \mu_{zz} = 4B\tau + \mu \end{aligned} \tag{7.5}$$

$$B = G\ell^2$$

For the sake of simplicity it has been assumed here that the additional constant $\eta = 0$. Inserting (7.4) into the equation of equilibrium we obtain the relations between the antisymmetric parts of σ_{ij} (symbol ∇^2 denotes here the two-dimensional Laplacean, $\nabla^2 = \frac{\partial^2}{\partial x^2} + \frac{\partial^2}{\partial y^2})$:

(7.6)
$$\overset{a}{\sigma}_{xy,y} + \overset{a}{\sigma}_{xz,z} = 0 \, , \quad \overset{a}{\sigma}_{xy,x} - \overset{a}{\sigma}_{yz,z} = 0$$
$$\overset{a}{\sigma}_{xz,x} + \overset{a}{\sigma}_{yz,y} = -G\tau\nabla^2\varphi .$$

From the remaining equations of equilibrium involving the couple-stresses and Eq. (7.5) stresses $\overset{a}{\sigma}_{ij}$ can be expressed in terms of μ and $\varphi(x,y)$,

(7.7)
$$\overset{a}{\sigma}_{xy} = -\frac{1}{2}\mu_{,z} ,$$
$$\overset{a}{\sigma}_{xz} = \frac{1}{2}\mu_{,y} - B\tau\frac{\partial\nabla^2\varphi}{\partial x} ,$$
$$\overset{a}{\sigma}_{yz} = -\frac{1}{2}\mu_{,x} - B\tau\frac{\partial\nabla^2\varphi}{\partial y} .$$

Substitution of (7.7) into (7.6) leads to the differential equation for $\varphi(x,y)$ sought for,

(7.8)
$$\nabla^2\varphi - \ell^2\nabla^2\nabla^2\varphi = 0 .$$

Taking into account that, due to the geometry of torsion, all stresses are independent of z, and hence $\mu_{,z} = 0$, the expressions for σ_{ij} take the form

$$\sigma_{xx} = \sigma_{yy} = \sigma_{zz} = \sigma_{xy} = \sigma_{yx} = 0$$

$$\begin{aligned}
\sigma_{xz} &= G\tau(\varphi_{,x} - y) + \frac{1}{2}\mu_{,y} - B\tau\nabla^2\varphi_{,x} \\
\sigma_{yz} &= G\tau(\varphi_{,y} + x) - \frac{1}{2}\mu_{,x} - B\tau\nabla^2\varphi_{,y} \\
\sigma_{zx} &= G\tau(\varphi_{,x} - y) - \frac{1}{2}\mu_{,y} + B\tau\nabla^2\varphi_{,x} \\
\sigma_{zy} &= G\tau(\varphi_{,y} + x) + \frac{1}{2}\mu_{,x} + B\tau\nabla^2\varphi_{,y} .
\end{aligned} \tag{7.9}$$

Expressions (7.8) and (7.5) satisfy the equations of equilibrium provided $\varphi(x,y)$ is a solution of differential equation (7.8).

The boundary conditions for φ are found from the conditions of vanishing of the surface loadings on the lateral surface of the bar. To this end let us establish at an arbitrary point A of the boundary curve of cross-section F a local Cartesian coordinate system (n, t, g) with axes coinciding with the normal, tangential and generator directions, respectively (Fig. 16). With $n_x = \cos\vartheta$, $n_y = \sin\vartheta$, $n_z = 0$ we have

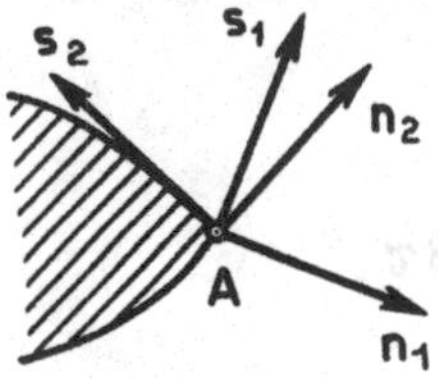

Fig. 16

$$\overset{n}{\mu} = 2B\tau\left(\frac{\partial^2\varphi}{\partial n\,\partial s} - 1\right) + \mu . \tag{7.10}$$

Five reduced boundary conditions read now (with $r^2 = x^2 + y^2$)

$$\bar{p}_x = \bar{p}_y = \bar{q}_z = 0 \qquad \text{(satisfied identically)}$$

$$\bar{p}_z = \sigma_{xz}\cos\vartheta + \sigma_{yz}\sin\vartheta - \frac{1}{2}\left(\overset{n}{\mu}_{,y}\cos\vartheta - \overset{n}{\mu}_{,x}\sin\vartheta\right) = 0,$$

(7.11)
$$\bar{q}_x = 2B\tau\frac{\partial^2\varphi}{\partial n^2}\sin\vartheta = 0,$$

$$\bar{q}_y = 2B\tau\frac{\partial^2\varphi}{\partial n^2}\cos\vartheta = 0,$$

Eqs. (4.11) can easily be reduced to the following two conditions

(7.12)
$$\frac{\partial}{\partial n}\left[\varphi - \ell^2\left(\nabla^2\varphi + \frac{\partial^2\varphi}{\partial s^2}\right)\right] = r\frac{\partial r}{\partial s},$$

$$\frac{\partial^2\varphi}{\partial n^2} = 0,$$

which have to be satisfied by $\varphi(x,y)$ at the boundary S of the cross-sectional area F.

If A were a singular point of the boundary, the resulting line load Q would equal $\frac{1}{2}(\overset{n+}{\mu} - \overset{n-}{\mu})$. Taking into account Eq. (7.10) we have

(7.93)
$$\bar{Q} = \left(\frac{\partial^2\varphi}{\partial n_1\partial s_1} - \frac{\partial^2\varphi}{\partial n_2\partial s_2}\right)2B\tau,$$

where n_i, s_i are indicated in Fig. 16. This phenomenon, similar to the existence of concentrated reactions arising in the corner of a plate in the classical plate theory, leads to certain unbalanced vertical forces. To avoid it we can assume the boundary curve to be of class C_1 (rounded corners).

To conclude the solution of the torsion problem one has to calculate the resulting forces acting at the free end of the cylinder.

Transversal forces vanish at the end section since we have

$$P_x = \iint_F \bar{p}_x \, dx \, dy + \int_S \bar{Q} \frac{\partial x}{\partial s} ds .$$

The second right-hand integral corresponds to the influence of the line force $\bar{Q}$ acting along the edge S. Writing the integrands in terms of stresses we obtain

$$P_x = \iint_F \left(\sigma_{zx} + \frac{1}{2} \frac{\partial \mu_{zz}}{\partial y}\right) dx \, dy + \frac{1}{2} \int_S (\mu_{zz} - \overset{n}{\mu}) \frac{\partial x}{\partial s} ds .$$

Inserting here the values of σ_{zx}, μ_{zz}, $\overset{n}{\mu}$ from Eqs.(7. 5), (7. 9), (7. 10) and making use of the Gauss-Green theorem we can write

$$P_x = \int_S \left\{ G\tau x \left[\frac{\partial(\varphi - \ell^2 \nabla^2 \varphi)}{\partial n} - \frac{\partial(r^2/2)}{\partial s} \right] + \right.$$

$$\left. + 2 B\tau \nabla^2 \varphi \frac{\partial x}{\partial n} + B\tau \frac{\partial^2 \varphi}{\partial n \partial s} \frac{\partial y}{\partial n} \right\} ds .$$

Substituting here condition (7.12) we finally arrive after some tedious transformations and rearrangements at the result

$$P_x = B\tau \int_S \left[\frac{\partial}{\partial s}\left(x\,\frac{\partial^2 \varphi}{\partial n\,\partial s}\right) - 2\,\frac{\partial^2 \varphi}{\partial s\,\partial y}\right] ds = 0 .$$

In a similar way it is shown that $P_y = 0$. The longitudinal stress resultant $P_z = 0$ since $\sigma_{zz} = 0$ everywhere.

The resulting twisting moment M_z is expressed by the integral

$$M_z = \iint_F (x\,\bar{p}_y - y\,\bar{p}_x + \bar{q}_z)\,dx\,dy + \int_S \bar{Q}\left(x\,\frac{\partial y}{\partial s} - y\,\frac{\partial x}{\partial s}\right) ds. \tag{7.14}$$

After some transformations, similar to those applied above, we arrive at the formula for M_z

$$M_z = G\tau \iint_F \left[x\,\frac{\partial(\varphi + \ell^2\nabla^2\varphi)}{\partial y} - y\,\frac{\partial(\varphi + \ell^2\nabla^2\varphi)}{\partial x} + x^2 + y^2 + 6\ell^2 \right] dx\,dy. \tag{7.15}$$

It is interesting to note that the same boundary conditions (7.12) for $\varphi\,(x, y)$ are obtained if we require the all six boundary tractions (instead of the five reduced ones) to vanish at the lateral surface of the bar (cf. Fig. 15) :

$$\sigma_{nn} = 0 \qquad \text{(satisfied identically)}$$

$$\sigma_{ns} = 0 \qquad \text{(satisfied identically)}$$

$$\mu_{nt} = 0 \qquad \text{(satisfied identically)}$$

$$\sigma_{nt} = G\tau \frac{\partial(\varphi - \ell^2 \nabla^2 \varphi)}{\partial n} + \frac{1}{2}\frac{\partial}{\partial s}(\mu - G\tau r^2) = 0, \qquad (7.16.1)$$

$$\mu_{nn} = 2B\tau\left(\frac{\partial^2\varphi}{\partial n\, \partial s} - 1\right) + \mu = 0, \qquad (7.16.2)$$

$$\mu_{ns} = -2B\tau \frac{\partial^2\varphi}{\partial n^2} = 0. \qquad (7.16.3)$$

Eliminating the indeterminate value of μ from (7.16.1) by means of (7.16.2) we are led to the conditions

$$\frac{\partial}{\partial n}\left[\varphi - \ell^2\left(\nabla^2\varphi + \frac{\partial^2\varphi}{\partial s^2}\right)\right] = \frac{\partial(r^2/2)}{\partial s},$$

$$\frac{\partial^2\varphi}{\partial n^2} = 0,$$

which coincide exactly with (7.12). The same applies to the calculation of the resultant forces acting in any of the transversal cross-sections of the bar. For instance, the twisting moment M_z can also be calculated from the expression

$$M_z = \iint_F (x\sigma_{zy} - y\sigma_{zx} + \mu_{zz})\,dx\,dy \,. \tag{7.17}$$

Both formulae (7.14), (7.17) lead to the same result enabling us to calculate the influence of the couple-stresses upon the torsional rigiditiy of the bar.

Passing with ℓ to the limit $\ell = 0$ the governing equation (8.11) of the entire problem as well as the expression for the torque (8.17) are reduced to the classical equations known from the elementary theory of torsion.

A particularly simple form take the derived equations for the couple-stress theory in the case when the cross-section of the cylinder becomes circular. Here, as in the classical theory, the warping function can be assumed to be identically zero. The most interesting result, as it was pointed out by Koiter [1] , is furnished by the formula (7.15) which gives the torsional rigidity of a circular bar. Performing the prescribed integration over the cross-sectional area of diameter d one is led to the result

$$J = \frac{\pi}{32} G d^4 \left[1 + 48\left(\frac{\ell}{d}\right)^2 (1+\eta)\right]$$

or

$$J = J_0\left[1 + 48\left(\frac{l}{d}\right)^2(1+\eta)\right], \qquad (7.18)$$

where J_0 denotes the traditional value of the torsional rigidity (without the couple-stresses). In contrast to our previous derivations, the additional elastic constant η has been taken into account in this formula. The simple formula (7.18) offers a certain possibility of experimental verification of the couple-stress theory: performing the torsion test on circular bars of various (possibly small) diameters and dividing the measured torsional rigidities by d^4, a series of numbers should be obtained. According to the classical theory the ratio J/d^4 equals $\pi G/32$ and is independent of the diameter. Formula (7.18) indicates however that the ratios should increase at decreasing values of d, especially for d being of the order of l.

8. Propagation of Surface Waves

A detailed study of the problem of wave propagation in micropolar media is not simple owing to the character of the wave equations (2.13), (2.15) describing the phenomenon. An interesting analysis of a problem of this type was discussed in paper [18] by C. Rymarz in 1967 ; it concerned the propagation of surface waves known in classical elasticity as Rayleigh waves.

The fundamental equations of the problem have the form

$$
(8.1)\qquad
\begin{aligned}
&\sigma_{ji,j} + \varrho X_i = \varrho \ddot{u}_i ,\\
&\mu_{ji,j} + \epsilon_{ijk}\,\sigma_{jk} + \varrho Y_i = 0 ,
\end{aligned}
$$

the assumptions concerning the equivalence of the partial and material time-derivatives being made. After reduction of the indeterminate invariant μ and the skew-symmetric components of the force-stress tensor Eqs. (8.1) are reduced to a single system of three equations of motion (2.19)

$$
(8.2)\qquad s_{ji,j} - \frac{1}{2}\,\epsilon_{kji}\left[m_{\ell k,\ell j} + \left(\varrho Y_k\right)_{,j}\right] + \varrho X_i = 0
$$

Using the stress-strain relations (2.39) and the geometric relations (2.40) these equations can be transformed to a system of three equations of motion expressed in displacements which are easily recognized as the generalization of the

Navier equations well-known from the classical elasticity

$$G u_{i,jj} + (\lambda + G) u_{j,ji} - G\ell^2 (u_{i,jj} - u_{j,ji})_{,kk} + \\ + \varrho\left(X_i - \frac{1}{2} \epsilon_{ijk} Y_{j,k}\right) = \varrho \ddot{u}_i . \tag{8.3}$$

If $\underline{Y} = 0$ and $\ell = 0$, then Eq. (8.3) assumes the familiar Navier form.

Let us recall from the classical elastokinetics that the Rayleigh waves are propagated along a free surface of the body and penetrate but little into the body, their amplitude decreasing exponentially with increasing depth in the solid. To simplify the problem let us assume that the body forms a semi-infinite half-space $x_3 > 0$ and that the wave is plane, independent of x_2, thus propagating along the x_1-axis (Fig. 17).

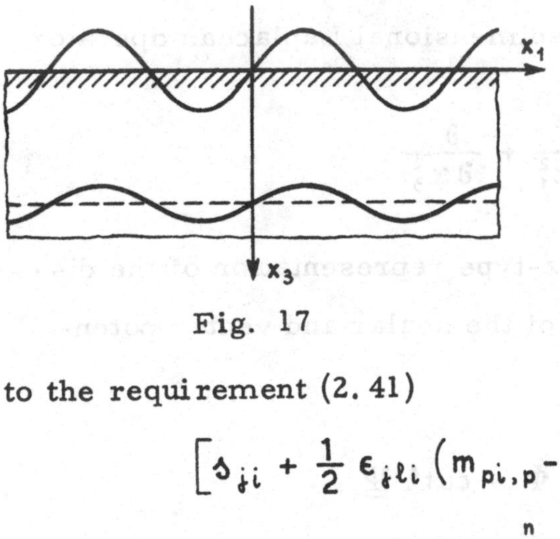

Fig. 17

The equilibrium equations being formulated in stresses, the boundary conditions are to be written in stresses, too. The condition of surface $x_3 = 0$ to be free of external loads reduces to the requirement (2.41)

$$\left[\sigma_{ji} + \frac{1}{2} \epsilon_{jli} \left(m_{pi,p} - \overset{n}{m}_{,i} + \varrho Y_l\right)\right] n_j = 0, \\ m_{ji}\, n_j - \overset{n}{m}\, n_i = 0 . \tag{8.4}$$

In the particular case under consideration, the equations of motion, owing to the plane state of strain (in x_1, x_3) - are considerably simplified. The simplification is also due to the fact that in the problem of surface waves propagation the body forces and couples are assumed to identically vanish throughout the body. Under these conditions the equations of motion (8. 3) can be explicitly written as

$$\begin{aligned} &G\nabla^2 u_1 + (\lambda + G)(u_{1,11} + u_{3,31}) - \\ &- G\ell^2\nabla^2(\nabla^2 u_1 - u_{1,11} - u_{3,31}) = \rho\ddot{u}_1, \end{aligned} \tag{8.5}$$

$$\begin{aligned} &G\nabla^2 u_2 + (\lambda + G)(u_{1,13} + u_{3,33}) - \\ &- G\ell^2\nabla^2(\nabla^2 u_3 - u_{1,13} - u_{3,33}) = \rho\ddot{u}_3, \end{aligned}$$

where ∇^2 denotes the two-dimensional Laplacean operator

$$\nabla^2 = \frac{\partial^2}{\partial x_1^2} + \frac{\partial^2}{\partial x_3^2} .$$

The usual Helmholtz-type representation of the displacement vector with the aid of the scalar and vector potentials Φ, $\underline{\Psi}$

$$\underline{u} = \operatorname{grad}\Phi + \operatorname{curl}\underline{\Psi}$$

proves again to be most useful. In the two-dimensional case this representation is reduced to

$$u_1 = \Phi_{,1} + \Psi_{,3} \quad , \quad u_3 = \Phi_{,3} - \Psi_{,1} \tag{8.6}$$

which, substituted into (8. 5), yields

$$\begin{aligned} &(\lambda + 2G)\nabla^2\Phi_{,1} + G\nabla^2\Psi_{,3} - \\ &- G\ell^2\nabla^2\nabla^2\Psi_{,3} = \varrho(\ddot{\Phi}_{,1} + \ddot{\Psi}_{,3}), \\ &(\lambda + 2G)\nabla^2\Phi_{,2} - G\nabla^2\Psi_{,1} + \\ &+ G\ell^2\nabla^2\nabla^2\Psi_{,1} = \varrho(\ddot{\Phi}_{,3} - \ddot{\Psi}_{,1}). \end{aligned} \tag{8.7}$$

A particular integral of (8. 7) is obtained by assuming that all terms involving Φ and Ψ vanish independently and by performing the respective integrations. We are thus led to a system of two separated wave-type equations for Φ and Ψ

$$\begin{aligned} &\nabla^2\Phi - \frac{1}{c_L^2}\ddot{\Phi} = 0 \\ &\nabla^2(1 - \ell^2\nabla^2)\Psi - \frac{1}{c_T^2}\ddot{\Psi} = 0, \end{aligned} \tag{8.8}$$

which can be considered as basic equations governing the problem of propagation of plane elastic waves in micropolar media. With $\ell \longrightarrow 0$ Eqs. (8. 8) become identical with the classical elastic wave equations; c_L and c_T denote the velocities of propagation of longitudinal and transversal waves, respectively,

(8.9) $$c_L^2 = \frac{\lambda + 2G}{\varrho}, \quad c_T^2 = \frac{G}{\varrho}.$$

It has to be mentioned here that C. Rymarz in his original paper [18] considers a more general thermo-elastic problem : the free energy expression contains terms depending on the temperature of the medium; owing to the assumed adiabatic character of the wave propagation process, the temperature is eliminated from the equation of motion the latter being reduced to wave equation analogous to (8. 8); the only difference consists in the Lamé constant λ which is replaced by

$$\lambda_s = \lambda + \frac{\beta}{c_\varepsilon T_0}$$

with $\beta\nu = \alpha(1+\nu)$, α denoting the thermal expansion coefficient, T_0 the absolute temperature of the reference state and c_ε - specific heat at constant strain. Apart from the slight variation of constants, the general character of solutions remains unchanged.

Following the usual way of derivation of the surface waves propagation, the potentials Φ , Ψ are assumed in the form

(8.10) $$\begin{aligned} \Phi &= A \exp\left[-\alpha x_3 + i(kx_1 - \omega t)\right], \\ \Psi &= B \exp\left[-\beta x_3 + i(kx_1 - \omega t)\right], \end{aligned}$$

where A, B are the amplitudes of vibration of the surface of the solid, α, β characterize the rate of decrease of amplitudes with increasing depth, ω - is the angular frequency of the motion, ω / k is the phase velocity of the wave propagation. Constants A, B are in general complex numbers whereas α, β should be real and positive in order to secure the exponential "damping" of vibrations under the surface $x_3 = 0$.

Equations (8.10) substituted into (8.8), the following relations between α, β, k and ω are obtained :

$$\alpha^2 = k^2 - \omega^2 / c_L^2$$

$$(\beta^2 - k^2)\left[1 - \ell^2(\beta^2 - k^2)\right] + \omega^2 / c_T^2 = 0 .$$

The second equation can be solved for β to give two possible values β_1, β_2,

$$\beta_{1,2}^2 = k^2 + \frac{1}{2\ell^2} \pm \frac{1}{2}\sqrt{\frac{1}{\ell^4} + \frac{4\omega^2}{c_T^2 \ell^2}} . \qquad (8.12)$$

Constants β_1, β_2 are to be real which leads to the requirement

$$k^2 + \frac{1}{2\ell^2} > \frac{1}{2}\sqrt{\frac{1}{\ell^4} + \frac{4\omega^2}{c_T^2 \ell^2}} ,$$

or

$$\frac{v}{c_T} < \sqrt{1 + \left(\frac{2\pi\ell}{\lambda}\right)^2} ,$$

where v denotes the phase velocity of surface waves and λ the wave length

$$v = \frac{\omega}{k}\,, \quad \lambda = \frac{2\pi}{k}\,.$$

From (8.11) it follows that α^2 is positive as long as

$$k^2 - \frac{\omega^2}{c_L^2} > 0 \quad \text{or} \quad \frac{v}{c_L} < 1\,.$$

Hence the first limitations for the phase velocity of the motion reduce to

$$\frac{v}{c_L} < 1\,,$$

$$\frac{v}{c_T} < \sqrt{1 + \left(\frac{2\pi l}{\lambda}\right)^2} \tag{8.13}$$

Recalling that the ratio

$$c_L / c_T = \sqrt{\frac{2(1-\nu)}{1-2\nu}}$$

depends on ν and varies from $\sqrt{2}$ to infinity as ν varies from 0 to 1/2, the region of existence of surface waves following from (8.13) may be plotted on the plane $(v/c_T\,,\ 2\pi l/\lambda)$, (Fig. 18). The shadowed area shows as an example, the region described by Eqs. (8.13) in the particular case $\nu = 0$.

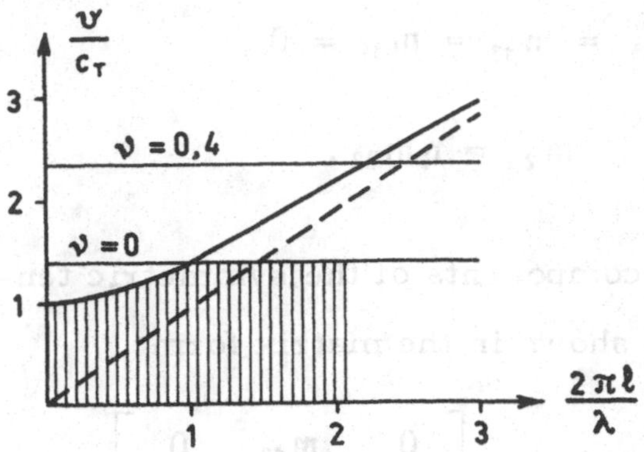

Fig. 18

In order to determine the remaining constants in (8.10) and possible additional limitations of the phase velocity, let us return to the boundary conditions on the free surface $x_3 = 0$ of the solid.

The boundary conditions (8.4) are considerably simplified in the particular case of plane strain and horizontal boundary surface $x_3 = 0$. Since $\underline{n} = \underline{n}(0,0,-1)$, Eqs. (8.4) read now

$$s_{3i} + \frac{1}{2}\epsilon_{3li}(m_{pl,p} - m_{33,l}) = 0,$$
$$m_{3i} - m_{33}\delta_{3i} = 0, \tag{8.14}$$

and the conditions of plane strain derived in Section 4 (in a different plane x_1, x_2, however) state that

$$u_1 = u_1(x_1, x_3), \quad u_2 = 0, \quad u_3 = u_3(x_1, x_3),$$

$$\omega_1 = \omega_3 = 0,$$

all $\varkappa_{ij}$ except $\varkappa_{21}$ and $\varkappa_{23}$ are zero;

$$s_{12} = s_{23} = 0, \quad s_{22} = \nu(s_{11} + s_{22}),$$

$$m_{11} = m_{22} = m_{33} = m_{13} = m_{31} = 0\,,$$

$$m_{21} = \eta m_{12}\,, \quad m_{23} = \eta m_{32}\,.$$

The non-vanishing components of the symmetric tensor σ_{ij} and deviator m_{ij} are shown in the matrix form

$$\sigma_{ij} = \begin{bmatrix} \sigma_{11} & 0 & \sigma_{13} \\ 0 & \sigma_{22} & 0 \\ \sigma_{13} & 0 & \sigma_{33} \end{bmatrix}, \quad m_{ij} = \begin{bmatrix} 0 & m_{12} & 0 \\ m_{21} & 0 & m_{23} \\ 0 & m_{32} & 0 \end{bmatrix}.$$

Eqs. (8.14) can thus be explicitly written as

$$\begin{aligned} &\sigma_{31} - \frac{1}{2}\left(m_{12,1} + m_{32,3}\right) = 0\,, \\ (8.15) \qquad & \\ &\sigma_{33} = m_{32} = 0\,. \end{aligned}$$

Now stresses σ_{31}, σ_{33}, m_{12}, m_{32} are to be expressed in displacements u_1, u_3, according to formulae (2.39), (2.40). Substituting now u_1, u_3 expressed in terms of Φ, Ψ according to (8.6) we are led to the following three relations

$$\begin{aligned} &(\lambda + 2G)\Phi_{,33} + \lambda\Phi_{,11} - 2G\Psi_{,13} = 0\,, \\ (8.16) \qquad &2\Phi_{,13} + \Psi_{,33} - \Phi_{,11} - \ell^2\nabla^2\nabla^2\Psi = 0\,, \\ &\nabla^2\Psi_{,3} = 0\,. \end{aligned}$$

It is to be observed that the constant η does not

enter these relations since it appears in expressions involving m_{21} and m_{23}, and these components are absent from the boundary conditions (8.15).

Expressions (8.10) are now substituted into (8.16); it is to be remembered here that owing to two alternate values of β (Eq. (8.12)) expression (8.10) for Ψ has to be written in the form

$$\Psi = \left[B_1 \exp(-\beta_1 x_3) + B_2 \exp(-\beta_2 x_3) \right] \exp i(kx_1 - \omega t).$$

The resulting equations (written for $x_3 = 0$) are

$$\begin{aligned} &\left[(\lambda + 2G)\alpha^2 - \lambda k^2 \right] A + 2Gik(B_1\beta_1 + B_2\beta_2) = 0, \\ &-2ik\alpha A + B_1\left[k^2 + \beta_1^2 - \ell^2(\beta_1^2 - k^2)^2 \right] + \\ &\quad + B_2\left[k^2 + \beta_2^2 - \ell^2(\beta_2^2 - k^2)^2 \right] = 0, \\ &B_1\beta_1(\beta_1^2 - k^2) + B_2\beta_2(\beta_2^2 - k^2) = 0. \end{aligned} \tag{8.17}$$

These are three homogeneous equations for the three constants A, B_1, B_2. In order to secure non-trivial solutions of Eqs. (8.17) the principal determinant of (8.17) has to vanish. After rather cumbersome and tedious transformations this condition is written in the form

$$\left[2-\left(\frac{v}{c_T}\right)^2\right]^4\left[\left(1+\frac{2\omega^2\ell^2}{c_T^2}\right)\right]\left(k^2\ell^2+\frac{1}{2}\right)+\frac{1}{2}+$$

(8.18)
$$+\frac{2\omega^2\ell^2}{c_T^2}+2\beta_1\beta_2\frac{\omega^2\ell^2}{c_T^2}\Big]=$$

$$=16\left(1+\frac{4\omega^2\ell^2}{c_T^2}\right)\left(1-\frac{v^2}{c_T^2}\right)\left[1+k^2\ell^2-\frac{\omega^2}{c_T^2k^2}\right].$$

Introducing here for the sake of brevity,the new variables used earlier for plotting the diagram in Fig. 18, namely

$$\xi=\frac{2\pi\ell}{\lambda}\ ,\quad \eta=\frac{v}{c_T}\ ,$$

remembering that $v/c_L=\eta c_T/c_L=\eta\gamma$ and using (8.11), (8.12), Eq. (8.18) is rewritten as

$$(2-\eta^2)^4\left[1+2\eta^2\xi^4+\xi^2+3\xi^2\eta^2+\right.$$

(8.19)
$$\left.+2\xi^2\eta^2\sqrt{\xi^2+1-\eta^2}\right]=16(1+4\xi^2\eta^2)(1-\gamma\eta^2)(1+\xi^2-\eta^2).$$

Eq. (8.19)gives the relationship between the surface waves velocity v represented by variable η and the elastic constants (ℓ represented by ξ and ν appearing in γ and η). Passing to the case of a classical elastic medium one has to put $\ell=0$ (and hence $\xi=0$) reducing thus (8.19) to

(8.20)
$$(2-\eta^2)^4=16(1-\gamma\eta^2)(1-\eta^2)$$

identical with the classical Rayleigh result (cf. e.g. [19]).

Two important features distinguishing the couple-stress case from the classical one can be noted here. The first one is easily recognized at the first sight of Eq. (8.19); in contrast to (8.20) the stress wave velocity v depends upon the wave length or the ratio l/λ entering the variable ξ, whereas in (8.20) the ratio v/c_T depends solely upon ν. Hence, the surface waves are dispersive when the couple-stresses are taken into consideration, which could be guessed from the form of equations (8.8) where the orders of the time and space derivatives do not agree.

The second important feature is observed when Eq. (8.19) is solved for η. The solution was found by C. Rymarz numerically and is represented in Fig. 19, where the curves $\eta = f(\xi)$ are superimposed on the regions of existence of solutions shown in Fig. 18. The classical, non-dispersive wave velocities ($l = 0$) for $\nu = 0$ and $\nu = 0{,}4$ are marked on the vertical axis. In the couple-stress case, increasing values of l rapidly accelerates the surface waves, their velocity c_s easily reaching and passing the limit c_T. However in the vicinity

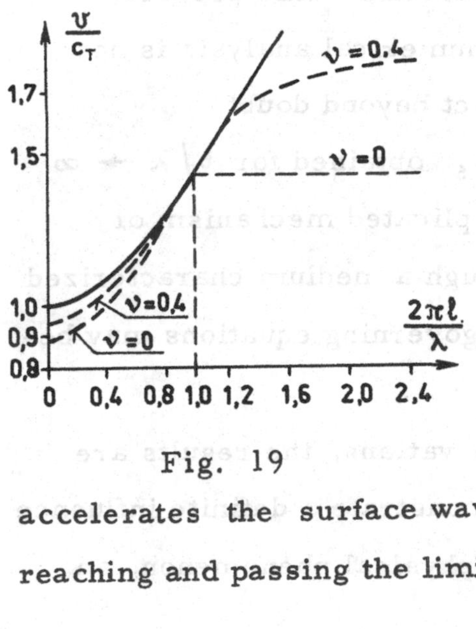

Fig. 19

of $\xi = 1$, i. e. for l being of the order of wave length λ, the curves seem to disappear from the region of validity of solutions and reappear for larger values of l. For still larger values of l/λ, c_s approaches asymptotically the longitudinal waves velocity (whereas in classical elasticity it does not exceed the value of $0,6\,c_L$).

To summarize these results we can observe the following properties of the derived solutions.

(a) A definite increase of the surface waves velocity can be observed even for very small values of l (and large wave lengths λ). The dispersive properties of the motion are very strong at $l = \lambda/12$ to $\lambda/6$.

(b) For certain values of l/λ in the vicinity of $l = \lambda/6$ - the solution does not exist at all and the surface waves cannot be propagated in the medium. This property seems to depend on ν - though the numerical analysis is not accurate enough to establish this fact beyond doubt.

(c) The large values of c_s obtained for $l/\lambda \rightarrow \infty$ are questionable in view of the complicated mechanism of transmitting very short waves through a medium characterized by a "coarse" microstructure; the governing equations may become inaccurate in these cases.

In spite of all these reservations, the results are certainly interesting since they demonstrate a definite influence of couple-stresses on an important physical phenomenon, an influence of a qualitative nature which makes the experimental verification of results more probable.

9. Statical Problem of a Rigid Punch

Another problem leading to infinite stress concentrations follows from the consideration of a rigid, smooth and flat-ended punch acting on the surface of an elastic body. The problem of such a punch of finite dimensions (in the plane state of strain) was considered by Muki and Sternberg ([2]) in 1965. Let us pass to more detailed discussion of a similar problem solved in [20] and concerning an infinite elastic strip compressed between two perfectly rigid blocks extending from $x = 0$ to plus infinity (Fig. 20). The problem is solved with

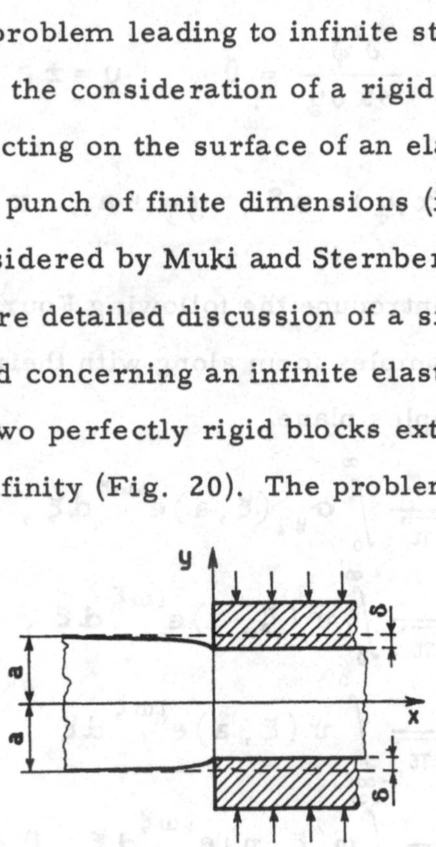

Fig. 20

the aid of the stress functions φ , ψ introduced in Sect. 4, Eqs. (4.19), (4.20).

The boundary conditions of the problem take the form

$$-\frac{\partial^2\varphi}{\partial x\partial y}+\frac{\partial^2\varphi}{\partial x^2}=0\,, \qquad y=\pm a$$

$$\frac{\partial\psi}{\partial y}=0\,, \qquad y=\pm a$$

(9.1)

$$\frac{\partial^2\varphi}{\partial x^2}+\frac{\partial^2\psi}{\partial x\partial y}=0\,, \qquad y=\pm a\,,\ x<0\,,$$

$$v(x,y)=\mp\delta\,, \quad y=\pm a\,, \quad x>0\,.$$

Let us introduce the following Fourier exponential transforms in a complex form along with their regions of existence in the complex plane

$$S^+(\omega)=\frac{a}{\sqrt{2\pi}}\int_0^\infty \sigma_{yy}(\xi,a)e^{i\omega\xi}d\xi\,, \quad \Im m\,\omega>0\,;$$

$$V^+(\omega)=\frac{a}{\sqrt{2\pi}}\int_0^\infty v(\xi,a)e^{i\omega\xi}d\xi\,, \quad \Im m\,\omega>0\,;$$

$$V^-(\omega)=\frac{a}{\sqrt{2\pi}}\int_{-\infty}^0 v(\xi,a)e^{i\omega\xi}d\xi\,, \quad \Im m\,\omega<\varepsilon_1\,;$$

$$\Phi(\omega)=\frac{a}{\sqrt{2\pi}}\int_{-\infty}^{+\infty}\varphi(\xi,\eta)e^{i\omega\xi}d\xi\,, \quad 0<\Im m\,\omega<\varepsilon_2\le\varepsilon_1;$$

$$\Psi(\omega)=\frac{a}{\sqrt{2\pi}}\int_{-\infty}^{+\infty}\psi(\xi,\eta)e^{i\omega\xi}d\xi\,, \quad 0<\Im m\,\omega<\varepsilon_2\le\varepsilon_1\,;$$

$$\eta=y/a\,.$$

Here $\varepsilon_1, \varepsilon_2$ denote certain positive constants which can be determined as a rule after the final solution of the problem.

Taking into account the symmetry of the problem, Eqs. (4.20) yield, after transformation, the following expressions for the stress functions

$$\Phi(w,\eta) = \frac{a}{w}\left[A \cosh w\eta + B w\eta \sinh w\eta\right]$$
$$\Psi(w,\eta) = C \sinh w\eta + D \sinh k\eta\,, \qquad (9.2)$$

where

$$k = \sqrt{(\ell^2 w^2 + a^2)/\ell^2}\,.$$

The transforms of stresses σ_{yy}, σ_{yx}, m_{yz} and displacement v can be expressed with the aid of (4.19) and (9.2) as follows

$$S_{yy}(w,\eta) = -\frac{w}{a}\left[A \operatorname{ch} w\eta + Bw\eta \operatorname{sh} w\eta\right] - \frac{iw}{a^2}\left[Cw\operatorname{ch} w\eta + Dk \operatorname{ch} k\eta\right],$$

$$S_{yx}(w,\eta) = \frac{iw}{a}\left[A \operatorname{sh} w\eta + B(\operatorname{sh} w\eta + w\eta \operatorname{ch} w\eta)\right] - \frac{w^2}{a^2}\left[C\operatorname{sh} w\eta + D\operatorname{sh} k\eta\right],$$

$$M_y(w,\eta) = \frac{1}{a}\left[Cw\operatorname{ch} w\eta + Dk\operatorname{ch} k\eta\right], \qquad (9.3)$$

$$V(w,\eta) = \frac{1}{2G}\left\{-A\operatorname{sh} w\eta + B\left[(1-2\nu)\operatorname{sh} w\eta - w\eta \operatorname{ch} w\eta\right]\right\} - \frac{wi}{2Ga}\left[C\operatorname{sh} w\eta + D\operatorname{sh} k\eta\right].$$

The constants of integration A, B, C, D are calculated from the boundary conditions (9.1) and the continuity conditions (4.20); the constants are expressed in terms of the unknown transform $S^+(w)$ of stress σ_{yy} on the boundary of the strip. The results are now substituted into the last boundary condition (9.1) to yield the following Wiener-Hopf equation

$$\frac{(1-\nu)a}{G} S^+(w) H(w) - V^+(w) = V^-(w) \tag{9.4}$$

where

$$H(w) = \frac{1}{w} \frac{a^2 k \, sh^2 w}{a^2 k (w + sh w \, ch w) + 4(1-\nu) w^2 \ell^2 p \, ch w},$$

$$V^+(w) = -\frac{1}{w} \frac{\delta a i}{\sqrt{2\pi}}, \tag{9.5}$$

$$p = k \, sh w - w \, ch w \, th k .$$

The kernel $H(w)$ in Eq. (9.5) with $\ell = 0$ (which corresponds to neglecting the couple-stress influence) transforms to

$$H(w) = \frac{1}{w} \frac{sh^2 w}{w + sh w \, ch w}$$

Eq. (9.4) becomes then identical with the equation obtained in [21].

Eq. (9.4) can approximately be solved with the aid of Koiter's method presented in [22] the complicated kernel (9.5) being replaced by a simple function $\bar{H}(w)$ fulfilling the three conditions :

(1) $H(w)$ and $\bar{H}(w)$ should behave similarly in the region of regularity of all transforms appearing in the problem (in our case, in the neighbourhood of the $Re(w)$-axis);

(2) $H(w)$ and $\bar{H}(w)$ should assume the same values in the neighbourhood of points $Re(w) = 0$ and $Re(w) = \infty$;

(3) $\bar{H}(w)$ should be characterized by a simple distribution of zeros and poles.

After satisfying these conditions, the approximate solutions $\bar{\sigma}_{yy}(x,a), v(x,a)$ do not considerably differ from the true results for arbitrary x and are accurate at $x \to 0$ and $x \to \pm\infty$.

Since from (9.5) it follows that

$$H(w) = \frac{1}{2} + O(w^1) \quad \text{when} \quad w \to 0$$

$$H(w) = \frac{1}{(3-2\nu)w} + O(w^{-2}) \quad \text{when} \quad w \to \infty$$

hence we assume

$$\bar{H}(w) = \frac{1}{(3-2\nu)\sqrt{w^2 + \frac{4}{(3-2\nu)^2}}} \tag{9.6}$$

Substituting $\overline{H}(w)$ from (9.6) into (9.4) we obtain an auxiliary Wiener-Hopf equation which is solved by factorization technique consisting in the separation of expressions analytic in complex half-planes $\operatorname{Im}(w) > 0$, $\operatorname{Im}(w) < \varepsilon$, respectively. If this procedure proves possible for a certain positive number ε, then the obtained results can be retransformed in the common strip of singularity $0 < \operatorname{Im}(w) < \varepsilon$.

Multiplying both sides of the auxiliary Wiener-Hopf equations by $\sqrt{w - 2i/(3-2\nu)}$ one obtains

$$\frac{(1-\nu)\, a\, S^{+}(w)}{G(3-2\nu)\sqrt{w + \frac{2i}{3-2\nu}}} + K\,\frac{\sqrt{w-2i/(3-2\nu)}}{w} =$$

$$= V^{-}(w)\sqrt{w - \frac{2i}{3-2\nu}}\,, \qquad K = \frac{\delta a i}{\sqrt{2\pi}}\,.$$

The term involving K can be decomposed into two parts analytic in the respective half-planes yielding

$$\frac{a(1-\nu)}{G(3-2\nu)\sqrt{w + \frac{2i}{3-2\nu}}}\, S^{+}(w) + \sqrt{-\frac{2i}{3-2\nu}}\,\frac{K}{w} =$$

(9.7)

$$= \sqrt{w - \frac{2i}{3-2\nu}}\; V^{-}(w) - \left(\sqrt{w - \frac{2i}{3-2\nu}} - \sqrt{-\frac{2i}{3-2\nu}}\right)\frac{K}{w}.$$

According to the Liouville theorem both sides of this equation represent an integral function of the type

$$a_0 + a_1 w + a_2 w^2 + \dots$$

Only the assumption that all coefficients $a_0, a_1, \dots$ are zero leads to stresses vanishing at infinity. Hence, Eq. (9.7) is equivalent to two equations

$$\frac{a(1-\nu)}{G(3-2\nu)\sqrt{w + \frac{2i}{3-2\nu}}} S^{+}(w) + \sqrt{-\frac{2i}{3-2\nu}}\,\frac{K}{w} = 0,$$

$$\sqrt{w - \frac{2i}{3-2\nu}}\, V^{-}(w) - \left(\sqrt{w - \frac{2i}{3-2\nu}} - \sqrt{-\frac{2i}{3-2\nu}}\right)\frac{K}{w} = 0 \tag{9.8}$$

Application of the inverse Fourier transforms leads to

$$\xi > 0, \quad \bar{\sigma}_{yy}(\xi, a) = -\frac{KG\sqrt{-2i}\,\sqrt{3-2\nu}}{(1-\nu)a^2\sqrt{2\pi}} \int_{\Gamma} \frac{\sqrt{w + \frac{2i}{3-2\nu}}}{w} e^{iw\xi}\, dw,$$

$$\xi < 0, \quad \bar{v}(\xi, a) = \frac{K}{\sqrt{2\pi}\, a} \int_{\Gamma} \left(1 - \frac{\sqrt{\frac{-2i}{3-2\nu}}}{\sqrt{w - \frac{2i}{3-2\nu}}}\right) \frac{e^{-iw\xi}}{w}\, dw. \tag{9.9}$$

The integrals in (9.9) are calculated by means of the method of residua. To calculate the first of the integrals in (9.9) the path of integration is closed in the lower half-plane, account being taken of the pole at $w = 0$ and the branchpoint at $w = \sqrt{-2i/(3-2\nu)}$. Finally, the approximate formula for the distribution of stresses at the boundary of the strip reads

$$(9.10) \qquad \bar{\sigma}_{yy}(\xi, a) = -\frac{2\delta G}{a(1-\nu)}\left(\sqrt{3-2\nu}\,\frac{e^{-\frac{2}{3-2\nu}\xi}}{\sqrt{2\pi\xi}} + \operatorname{erf}\sqrt{\frac{2\xi}{3-2\nu}}\right).$$

The solution of an analogous problem without the couple-stresses being taken into consideration (classical elastic body) has the form [23]

$$(9.11) \qquad \sigma_{yy}(\xi, a) = -\frac{2\delta G}{a(1-\nu)}\left(\frac{e^{-2\xi}}{\sqrt{2\pi\xi}} + \operatorname{erf}\sqrt{2\xi}\right).$$

It is seen that these formulae differ by the coefficient $\sqrt{3-2\nu}$ influencing the stress concentration at $x = 0$. In addition, it is observed that $\bar{\sigma}_{yy}(\xi, a)$ does not explicitly depend upon the characteristic length ℓ and no limiting procedure of $\ell \rightarrow 0$ can transform (9.10) into (911). A similar conclusion was drawn by Muki and Sternberg in paper [2] where, by using a different technique, the problem of a finite punch acting on the surface of a halfplane has been solved (the coefficient $3 - 2\nu$ appears in that paper too).

The comparison between the couple-stress and classical results is demonstrated in Fig. 21. Solid lines correspond to classical results (9.11), dotted lines correspond to couple-stress results (9.10). Two different values of Poisson's ratio are considered. The increase of the stress concentration factor is 41-73% and, even at a distance of $x = a/2$

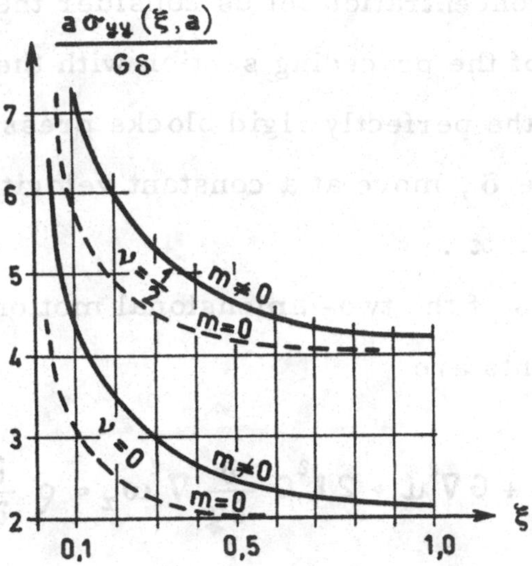

Fig. 21

from the singular point, 10-20% increase of normal stresses can be expected; the latter result is however a very rough estimate and certainly depends upon the assumed value of ℓ .

10. Dynamic Problem of a Moving Punch

As an example of a dynamic, time-dependent problem of infinite stress concentration let us consider the situation shown in Fig. 20 of the preceding section with the additional assumption that the perfectly rigid blocks pressed into the strip at the distance δ, move at a constant velocity c in the positive direction of x.

The equations of the two-dimensional motion written in terms of displacements are

$$(10.1)\qquad \begin{aligned} (\lambda + G)\frac{\partial e}{\partial x} + G\nabla^2 u + 2\ell^2 G \frac{\partial}{\partial y}\nabla^2\omega_z &= \varrho\frac{\partial^2 u}{\partial t^2}, \\ (\lambda + G)\frac{\partial e}{\partial y} + G\nabla^2 v - 2\ell^2 G \frac{\partial}{\partial x}\nabla^2\omega_z &= \varrho\frac{\partial^2 v}{\partial t^2}, \end{aligned}$$

where $e = \partial u/\partial x + \partial v/\partial y$ denotes the dilatation and ω_z is the rotation about the z-axis.

Introducing, as before, potentials of displacement $\varphi(x, y, t)$ and $\psi(x, y, t)$

$$u = \frac{\partial\varphi}{\partial x} + \frac{\partial\psi}{\partial y}, \qquad v = \frac{\partial\varphi}{\partial y} - \frac{\partial\psi}{\partial x},$$

$$e = \nabla^2\varphi, \quad \omega_z = -\frac{1}{2}\nabla^2\psi,$$

Eqs. (10.1) can be replaced by the system

$$(\lambda + 2G)\nabla^2\varphi = \varrho\ddot{\varphi}\,, \quad G\nabla^2\psi - \ell^2 G\nabla^4\psi = \varrho\ddot{\psi}\,. \tag{10.2}$$

The boundary conditions are the following

$$\begin{aligned}
\sigma_{yx}(x, \pm a) &= G\left(\frac{\partial^2\varphi}{\partial y^2} - 2\frac{\partial^2\varphi}{\partial x\,\partial y} - \frac{\partial^2\psi}{\partial x^2}\right) - \ell^2 G\nabla^4\psi = 0, \\
m_{yz}(x, \pm a) &= -2\ell^2 G\frac{\partial}{\partial y}\nabla^2\psi = 0. \\
\sigma_{yy}(x, \pm a) &= \lambda\nabla^2\varphi + 2G\frac{\partial}{\partial y}\left(\frac{\partial\varphi}{\partial y} - \frac{\partial\psi}{\partial x}\right) = 0, \quad x < ct, \\
v(x, \pm a) &= \frac{\partial\varphi}{\partial y} - \frac{\partial\psi}{\partial x} = \mp\delta, \qquad x > ct.
\end{aligned} \tag{10.3}$$

It is assumed that the process started at $t = -\infty$ and at the instant $t = 0$ the blocks are in the position indicated in Fig. 20.

In order to eliminate the time variable t let us introduce a new auxiliary variable $x' = x - ct$. Let

$$c_1 = \sqrt{\frac{\lambda + 2G}{\varrho}}$$

denote the longitudinal waves velocity and

$$c_2 = \sqrt{\frac{G}{\varrho}}$$

the velocity of transversal elastic waves; with the additional notations

$$\gamma^2 = 1 - \left(\frac{c}{c_1}\right)^2, \quad \beta^2 = 1 - \left(\frac{c}{c_2}\right)^2$$

Eqs. (10. 3) take the form

$$(10.4) \qquad \begin{aligned} &\gamma^2 \varphi_{,x'x'} + \psi_{,yy} = 0 \, , \\ &\beta^2 \psi_{,x'x'} + \psi_{,yy} - \ell^2 (\psi_{,x'x'x'x'} + 2\psi_{,x'x'yy} + \psi_{,yyyy}) = 0 \, . \end{aligned}$$

It will be assumed in the sequel that both β^2 and γ^2 are always positive; hance, the velocity c does not exceed the velocity of propagation of elastic waves in the medium.

Using the Fourier transforms introduced in Sect. 9, the solutions of the system (10. 4) - taking into account the symmetry of the problem - can be written as

$$\Phi(w, \eta) = A \operatorname{ch} \gamma w \eta \, , \quad \Psi(w, \eta) = B \operatorname{sh} s w \eta + C \operatorname{sh} p w \eta \, .$$

Here w is the transform parameter, $\eta = y/a$ and, moreover,

$$(10.5) \qquad \begin{aligned} &s = \sqrt{1 + \frac{1 + \sqrt{\Delta}}{2w^2 \lambda^2}} \, , \qquad p = \sqrt{1 + \frac{1 - \sqrt{\Delta}}{2w^2 \lambda^2}} \\ &\Delta = 1 + 4w^2 \lambda^2 (1 - \beta^2) \qquad \lambda = \frac{\ell}{a} \, . \end{aligned}$$

Using, as in Sect. 9, the boundary conditions (10. 3) the problem is reduced to the following Wiener-Hopf equation

$$\frac{a}{G} S^{+}(w) H(w) + \frac{M}{w} = V^{-}(w), \quad M = \frac{\delta a i}{\sqrt{2\pi}}. \tag{10.6}$$

The kernel of this equation has in our case a very complicated form

$$H(w) = \frac{\gamma(1-\beta^2) f(w)}{w\left[8\gamma\sqrt{\Delta}\sqrt{1+\frac{\beta^2}{w^2\lambda^2}} - (1+\beta^2)^2 f(w) \operatorname{cth} w\gamma\right]}, \tag{10.7}$$

where

$$f(w) = s\left(\sqrt{\Delta}+1\right) \operatorname{th} wp + p\left(\sqrt{\Delta}-1\right) \operatorname{th} ws.$$

It is easily verified that passing in (10 7) to the limit $\lambda \to 0$ one obtains

$$H(w) = \frac{\gamma(1-\beta^2) \operatorname{th}\gamma w \operatorname{th}\beta w}{w\left[4\beta\gamma \operatorname{th}\gamma w - (1+\beta^2)^2 \operatorname{th}\beta w\right]}. \tag{10.8}$$

This expression is identical with that obtained in [23] in which an analogous problem without the couple stresses has been considered.

In order to replaced Eq. (10.6) by the simplified equation and to apply Koiter's method, the behaviour of expression (10.7) in the interval $0<w<\infty$ (the function is symmetric) has to be discussed first, and in particular - the values of $H(w)$ at $w=0$ and $w=\infty$ should be determined. To this end function $H(w)$ is expressed in another form: using a new variable $u = w\lambda$ and introducing the notation

$$G(u) = \frac{\sqrt{1+\frac{\beta^2}{u^2}}\sqrt{\Delta} \operatorname{th}\frac{\gamma u}{\lambda}}{s(u)\left(\sqrt{\Delta}+1\right) \operatorname{th}\frac{p(u)u}{\lambda} + p(u)\left(\sqrt{\Delta}-1\right) \operatorname{th}\frac{s(u)u}{\lambda}}, \tag{10.9}$$

formula (10.7) is written as

$$(10.10) \quad H(u) = \frac{\gamma\lambda(1-\beta^2)}{(1+\beta^2)^2} \frac{\frac{1}{u}\,\mathrm{th}\,\frac{\gamma u}{\lambda}}{\frac{8\gamma}{(1+\beta^2)^2}\,G(u) - 1} .$$

Functions $p(w,\lambda)$, $s(w,\lambda)$, $\Delta(w,\lambda)$ introduced before are now functions of the single variable $u = w\lambda$, $\lambda \neq 0$, namely

$$p(u) = \sqrt{1 + \frac{1-\sqrt{\Delta}}{2u^2}} , \quad s(u) = \sqrt{1 + \frac{1+\sqrt{\Delta}}{2u^2}} , \quad \Delta = 1+4(1-\beta^2)u^2.$$

Function $H(u)$ can be replaced by a simple expression of the type

$$(10.11) \qquad \bar{H}(u) = \frac{C_1}{\sqrt{C_2^2 + u^2}}$$

similar to (9.10) provided the denominator of (10.10) remains positive along the entire Re(u)-axis. If for a certain $u = u_0$ were

$$G(u_0) = \frac{(1+\beta^2)^2}{8\gamma}$$

then expression (10.11) should be replaced by

$$\bar{h}(u) = \frac{\sqrt{C_1^2 + C_2^2 w^2}}{(w_0^2 - w^2)} .$$

In paper [23] it has been established that such an assumption leads to non-integrable solutions; the common strip of regularity mentioned in Sect. 9 ceases to exist and a resonance with surface Rayleigh-type waves occurs.

In paper [23] dealing with couple-stress-free medium such a resonance was encountered when the velocity of the rigid blocks reached the Rayleigh waves velocity c_R only. In the case under consideration with couple-stresses being taken into account, the critical velocity depends upon the additional parameter λ; influence of this kind was mentioned before in connection with the paper [18] by Rymarz.

In order to investigate this phenomenon let us discuss the behavior of $G(u)$ in interval $(0, \infty)$ This function assumes the following limiting values at $u \to 0$ and $u \to \infty$:

$$\lim_{u \to 0} G(u) = \frac{\gamma}{2}, \quad \lim_{u \to 0} H(u) = \frac{\gamma^2(1-\beta^2)}{4\gamma^2-(1+\beta^2)^2} \tag{10.12}$$

and

$$\lim_{u \to \infty} G(u) = \frac{1}{2}, \quad \lim_{u \to \infty} H(u) = \frac{\lambda\gamma(1-\beta^2)}{u\left[4\gamma-(1+\beta^2)^2\right]} \tag{10.13}$$

Comparing these values with the results derived in [23] it is found that the limits $\lim\limits_{w \to \infty} H$ coincide whereas at $w \to \infty$ the ratio

$$\frac{[\lim H(w)]_{m\neq 0}}{[\lim H(w)]_{m=0}} = \frac{4\beta\gamma-(1+\beta^2)^2}{4\gamma-(1+\beta^2)^2} = F(c) \tag{(10.14)}$$

is different from unity. In the static case $c = 0$

$$F(0) = \frac{1}{3-2\nu}$$

in accordance with the results of the preceding section and the paper [21]. This ratio is independent of λ and depends solely upon the ratio of velocity c to the velocity of propagation of longitudinal and transversal waves and upon Poisson's ratio ν. At increasing velocity c this ratio decreases to 0 when $c \to c_R$.

The limits of $G(u)$ at $u \to 0$ and $u \to \infty$ determined by Eqs. (10.12), (10.13) enable us to establish that the denominator of (10.10) remains positive at these points since, with $0 < c < c_2$:

$$\frac{8\gamma}{(1+\beta^2)^2}\, G(0) = \frac{4\gamma^2}{(1+\beta^2)^2} > 1 ,$$

$$\frac{8\gamma}{(1+\beta^2)^2}\, G(\infty) = \frac{4\gamma}{(1+\beta^2)^2} > 1 .$$

In Fig. 22 the graph of function

$$g(u) = \frac{8\gamma}{(1+\beta^2)^2}\, G(u) \tag{10.15}$$

has been shown (in a semi-logarithmic scale) in the interval $(0, \infty)$ for three cases of velocity c : smaller, equal and exceeding the Rayleigh waves velocity c_R. In each of these cases three values of parameter $\lambda = 0{,}001$, $\lambda = 0{,}01$ and $\lambda = 0{,}1$ are taken into account. From the graphs it evidently follows that as long as $c < c_R$, function $g(u)$ is greater than unity and no resonance with surface waves can occur. When $c > c_R$, parameter λ can be chosen small enough to lead to a resonance; at $c = c_R$ the resonace occurs only if $\lambda \to 0$. This observation is confirmed by a simple analysis of formula (10.9) : at sufficiently small values of $u \ll 1$ and $\lambda \to 0$ the following approximate relations hold

$$\sqrt{\Delta} + 1 \approx 2\,, \quad \sqrt{\Delta} - 1 \approx 2(1-\beta^2)u^2\,,$$

$$\sqrt{1+\frac{\beta^2}{u^2}} \approx \frac{\beta}{u}\,, \quad p(u) \approx \beta\,, \quad s(u) \approx \frac{1}{u}\,,$$

$$\operatorname{th}\frac{su}{\lambda} \approx \operatorname{th}\frac{pu}{\lambda} \approx \frac{\gamma u}{\lambda} \approx 1\,,$$

from which it follows that the limiting value of $G(u)$ is

$$G(u) \approx \frac{\frac{\beta}{u}}{\frac{2}{u} + 2\beta(1-\beta^2)u^2} \approx \frac{\beta}{2}\,.$$

At $\lambda \to 0$ and $u \ll 1$ the denominator of (10.10) attains thus its minimum value

$$\frac{4\beta\gamma}{(1+\beta^2)^2} - 1$$

which, as it is known, is always positive for $c < c_R$, since equation

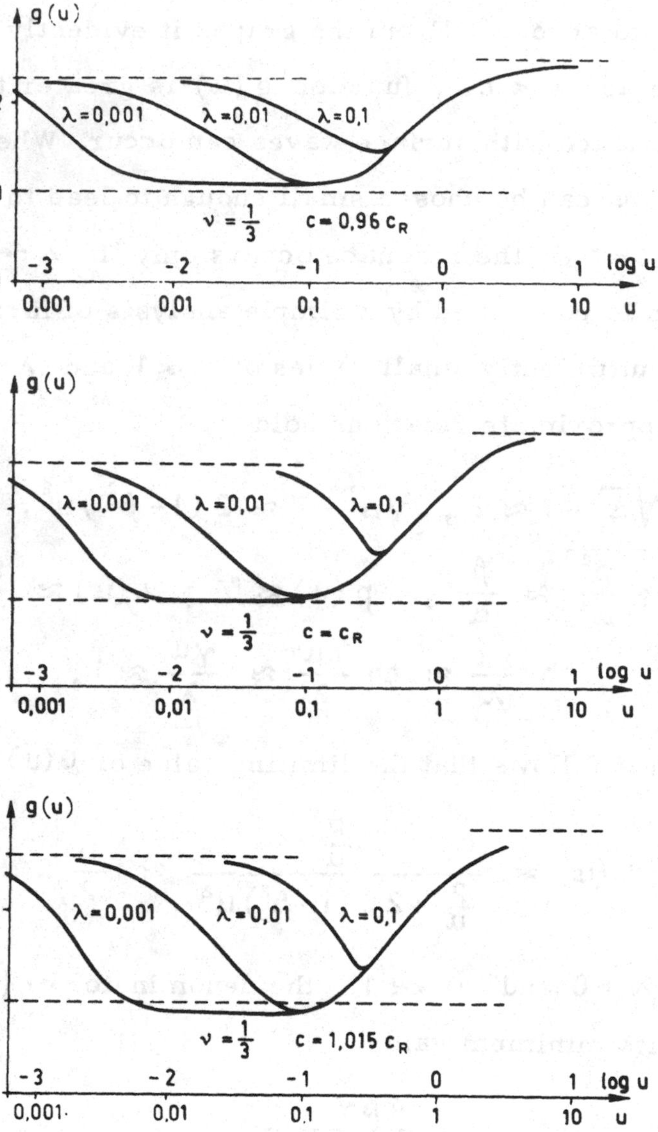
g(u)
λ=0,001
λ=0,01
λ=0,1
ν = 1/3
c = 0,96 c_R
-3
-2
-1
0
1
log u
0,001
0,01
0,1
1
10
u
g(u)
λ=0,001
λ=0,01
λ=0,1
ν = 1/3
c = c_R
-3
-2
-1
0
1
log u
0,001
0,01
0,1
1
10
u
g(u)
λ=0,001
λ=0,01
λ=0,1
ν = 1/3
c = 1,015 c_R
-3
-2
-1
0
1
log u
0,001
0,01
0,1
1
10
u

Fig. 22

$$4\beta\gamma - (1 + \beta^2)^2 = 0$$

is the well-known equation determining the velocity of propagation of the Rayleigh waves along the surface of an elastic body [19].

From Fig. 22 and the analysis of the function(10. 9) it is seen that the couple-stresses increase the critical velocity of the rigid blocks ; this increase may become considerable at great values of the parameter λ.

Under sufficiently small values of $c < c_R$ the denominator of (10. 10) - and hence (10. 7) - is positive for any real values of w. Taking into account the limiting values of $G(u)$ the kernel $H(w)$ can be replaced, as in Sect. 9, by the approximate expression

$$\bar{H}(w) = \frac{B}{\sqrt{w^2 + C^2}} \tag{10.16}$$

where

$$A = \frac{\gamma^2(1-\beta^2)}{4\gamma^2 - (1+\beta^2)^2}, \quad B = \frac{\gamma(1-\beta^2)}{4\gamma - (1+\beta^2)^2}, \quad C = \frac{B}{A}.$$

Performing the factorization in a way similar to that of the preceding section, the following formula for the approximate distribution of stresses under the rigid blocks is obtained

(10.17) $$\bar{\sigma}_{yy}(\xi,a) = -\frac{G\delta}{aA}\left(\frac{e^{-c\xi}}{\sqrt{\pi c \xi}} + \operatorname{erf}\sqrt{c\xi}\right)$$

where $\xi = x'/a$.

Formula (10.17), as before, is accurate if $\xi \to 0$ and $\xi \to \infty$. Hence, the rigorous value of the σ_{yy}-stress concentration coefficient at points $y = \pm a$, $\xi = 0$ is

(10.18) $$N' = \frac{G\delta}{a}\,\frac{\sqrt{\left[4\gamma-(1+\beta^2)^2\right]\left[4\gamma^2-(1+\beta^2)^2\right]}}{\gamma(1-\beta^2)\sqrt{\pi\gamma}}.$$

In Sect. 9 the stress concentration coefficient

(10.19) $$N = \frac{2G\delta}{a(1-\nu)}\,\frac{\sqrt{3-2\nu}}{\sqrt{2\pi}}$$

was obtained. It is easily verified that if $c \to 0$, and thus $\beta \to 1$ and $\gamma \to 1$, formula (10.18) is reduced to (10.19).

Fig. 23 demonstrates the variation of $\bar{\sigma}_{yy}(\xi, a)$ according to formula (10.17) for $\nu = 1/2$ and several values of c/c_2. It can be observed that here, similarly to the case when the couple-stresses are neglected (see paper [20]) the stresses $\bar{\sigma}_{yy}(\xi, a)$ decrease at increasing velocity c.

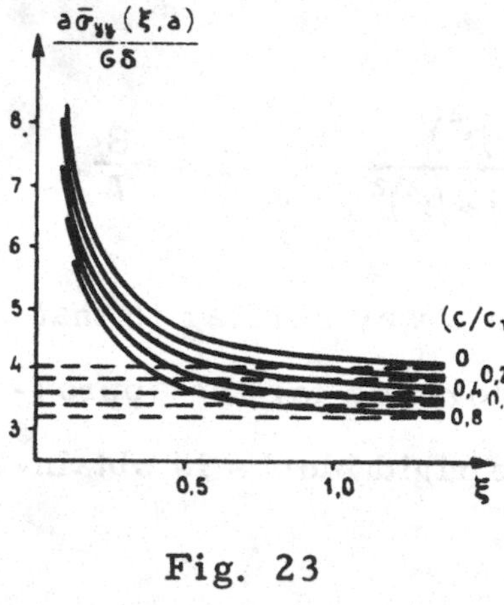

Fig. 23

The diagram of stresses $\sigma_{yy}(\xi, a)$ for $\nu = 1/2$ and two values of $c = 0$ and $c = 0{,}875\, c_2$ is shown in Fig. 24. Solid lines correspond to Eq. (10.17), dotted lines refer to the classical couple-stress-free solution given in[20]. Here, again, a certain increase of the stress and its concentration factor to the couple-stresses can be observed.

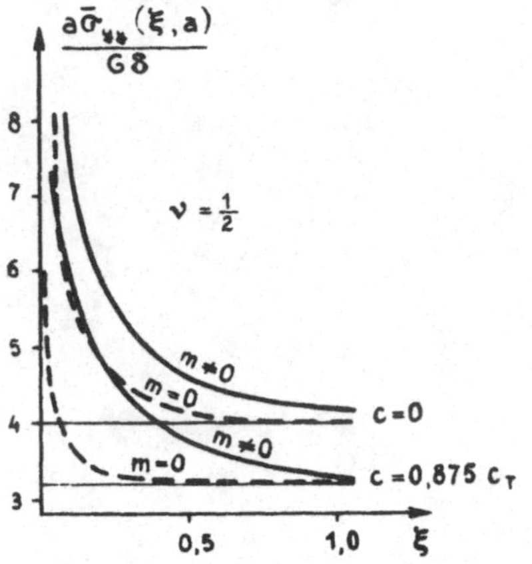

Fig. 24

This increase is the greatest at small velocities $c \to 0$ and equals then $\sqrt{3 - 2\nu}$; at increasing velocities $c \to c_R$ the analysis of Eqs. (10.18), (10.19) indicates that the ratio N'/N slightly diminishes.

References

[1] Koiter, W. T. : "Couple-Stresses in the Theory of Elasticity"; I-II, Proc. Kon. Ned. Akad. Wet. (1964), Ser. B. 67, No. 1 (17-44);

[2] Muki, R. ; Sternberg, E. : "The Influence of Couple-Stresses on Singular Stress Concentrations in Elastic Solids"; Z. A. M. P. (1965), Vol. 16, Fasc. 5 (611-648);

[3] Bogy, D. B. ; Sternberg, E. : "The Effect of Couple-Stresses on the Corner Singularity Due to an Asymmetric Shear Loading"; Int. J. Solids Structures (1968), Vol. 4 (159-174);

[4] Doyle, J. M. : "Singular Solutions in Elasticity"; Acta Mechanica (1966), Vol. IV, No. 1 (27-33);

[5] Morse, P. M. ; Feshbach, H. : "Methods of Theoretical Physics"; McGraw-Hill, New York-Toronto-London (1953);

[6] Mindlin, R. D. : "Influence of Couple-Stresses on Stress Concentrations"; Experimental Mechanics, January 1963, No. 1 (1-7);

[7] Sneddon, I. N. : "Fourier Transforms"; McGraw-Hill, New York (1951);

[8] Girkmann, K. : "Flächentragwerke"; Springer, Wien (1954);

[9] Timoshenko, S. ; Goodier, J. N. : "Theory of Elasticity"; McGraw-Hill, New York (1951);

[10] Banks, C. B.; Sokolówski, M. : "On Certain Two-Dimensional Applications of Couple-Stress Theory"; Int. J. Solids Structures (1968), Vol. 4 (15-29);

[11] Goodier, J. N. : "Concentration of Stress Around Spherical and Cylindrical Inclusions and Flaws"; Trans. Am. Soc. Mech. Engrs. (1933), AMP 55-7 (39-44);

[12] Muschelishvili, I. N. : "Certain Fundamental Problems of the Mathematical Theory of Elasticity" (in Russian); Moscow-Leningrad (1949);

[13] Savin, G. N. : "Stress Distribution Around Holes" (in Russian); Naukova Dumka, Kiev (1968);

[14] Savin, G. N. : "Foundations of Couple-Stress Theory" (in Russian), Kiev (1965);

[15] Nemish, Yu. N. : "Plane Couple-Stress Problem for Regions with Circular Hole" (in Russian); Prikl. Mekh. (1965), Vol. 1, No. 5;

[16] Savin, G. N.; Guz, A. N. : " On a Certain Method of Solution of Plane Couple-stress Problems in Multiply-Connected Regions" (in Russian); Prikl. Mekh. (1966), Vol. 2, No. 1;

[17] Sokolowski, M. : "Couple-Stresses in Problems of Torsion of Prismatic Bars"; Bull. Acad. Polon. Sci., Série Sci. Techn. (1965), Vol. 13, No. 8 (419-424);

[18] Rymarz, C. : "Surface Waves in a Medium with Couple-Stresses" (in Polish); Mech. Teoret. Stos. (1967), Vol. 5, No. 3 (337-346);

[19] Ewing, W.M.; Jardetzky, W.S.; Press, F. : "Elastic Waves in Layered Media"; McGraw-Hill, New York-Toronto-London (1957);

[20] Dusza, A.; Sokolowski, M. : "Couple-Stresses in the Discontinuous Boundary Value Problem of an Elastic Layer" (in Polish); Rozpr. Inzyn. (1968), Vol. 16, No. 1 (3-19);

[21] Sokolowski, M. : "Stresses in a Rigidly Clamped Plate Strip"; Arch. Mech. Stos. (1962), Vol. 14, No. 2 (271-283);

[22] Koiter, W.T. : "Approximate Solutions of Wiener-Hopf Type Integral Equations with Applications"; Proc. Kon. Ned. Akad. Wet. (1954), 57, No. 4;

[23] Matczynski, M.; Sokolowski, M. : "Quasi-Static Problem of a Rigidly Clamped Elastic Layer"; Arch. Mech. Stos. (1967), Vol. 19, No. 6 (867-882).

Contents

Printed in the United States
By Bookmasters